Airplane Performance on Grass Airfields

Airplane Performance on Grass Airfields presents an experiment-based approach to analysis and flight testing of airfield performance on grass runways. It discusses improvements for operational efficiency and safety of these airfields.

The book analyzes the interaction between the landing gear wheels and the surface of a grass runways during both takeoff and landing. Considering the ground performance of an aircraft on a grass runway, the book covers test methods and devices for measuring performance and introduces an information system for the surface condition of grass airfields: GARFIELD. The system is based on a tire–grass interaction model and uses digital soil maps, as well as current meteorological data obtained from a weather server.

The book is intended for researchers and practicing engineers in the fields of aviation and aircraft safety and performance.

Airplane Performance on Grass Airfields

Jaroslaw A. Pytka

CRC Press is an imprint of the
Taylor & Francis Group, an **informa** business

First edition published 2023
by CRC Press
6000 Broken Sound Parkway NW, Suite 300, Boca Raton, FL 33487–2742

and by CRC Press
4 Park Square, Milton Park, Abingdon, Oxon, OX14 4RN

CRC Press is an imprint of Taylor & Francis Group, LLC

Library of Congress Cataloging-in-Publication Data
Names: Pytka, Jaroslaw A., author.
Title: Airplane performance on grass airfields / Jaroslaw A. Pytka.
Description: First edition. | Boca Raton: CRC Press, [2023] | Includes bibliographical references and index.
Identifiers: LCCN 2022053427 | ISBN 9781032320786 (hbk) | ISBN 9781032320809 (pbk) | ISBN 9781003312765 (ebk)
Subjects: LCSH: Airplanes—Landing gear. | Private planes—Performance. | Airplanes—Takeoff. | Airplanes—Landing. | Runways (Aeronautics)—Testing. | Tires—Traction.
Classification: LCC TL682 .P98 2023 | DDC 629.133/34—dc23/eng/20230109
LC record available at https://lccn.loc.gov/2022053427

ISBN: 978-1-032-32078-6 (hbk)
ISBN: 978-1-032-32080-9 (pbk)
ISBN: 978-1-003-31276-5 (ebk)

DOI: 10.1201/9781003312765

Typeset in Times
by Apex CoVantage, LLC

Contents

Preface

This book was written as a reference in a niche sub-discipline that combines two areas of knowledge and technology, namely aeronautical engineering and terramechanics. I set myself the task to develop a practical application of methods known from terramechanics as well as new solutions in order to develop measuring devices and test procedures practically useful in the study of airfield performance of airplanes on a grass runway. Readers – aeronautical engineers, instrumentation design engineers, researchers, and graduate students as well as aviators – will find both the descriptions of devices and measurement methods as well as the results of tests that I carried out in real conditions, with airplanes, useful. I think that one of the most important advantages of this work are the results of the experiments, which can be both reference data and inspiration for the readers' own research. I've tried to keep the text simple and concise throughout. I've assumed that the potential readers have a background knowledge of aeronautical engineering, elementary soil mechanics, electronics, and metrology.

I express gratitude to several people, whose help was substantial. To Professor Zbigniew Pater, Rector of the Lublin University of Technology, who supported my research both good advice and financially; to Mr. Paweł Tomiło, my PhD student, who made some of my ideas a reality, in particular the design and construction of a measuring device with artificial intelligence; to pilots who carried out test aircraft flight tests, in particular to Mr. Krysztof Janusz from the Świdnik Aeroclub, for his excellent cooperation in the cockpit of the Koliber aircraft. Also, many thanks to my family members who helped with the research, in particular my eldest son Jan and his wife Kinga and my younger son Franciszek.

Lublin, October 2022

Preface

Acknowledgments

This publication has been financed from the Polish state budget under the program of the Minister of Education and Science "Polish Metrology" and project "Method and device for measuring airplane take-off and landing distance" PM/SP/0065/2021/12, with financing amount PLN 264.715 PLN and total project value PLN 264.715 PLN.

Author biography

Jaroslaw Alexander Pytka (DSc, PhD, MS, Automotive Engineer), born in 1968 in Wrocław, Poland, is working at the Department of Motor Vehicles, Lublin University of Technology, Poland, as Researcher and Educator for undergraduate and graduate students of Automotive Technology. He received his MS degree from the Lublin University of Technology in Automotive Engineering in 1992. He earned his PhD in Soil Physics from the Institute of Agrophysics, Lublin, Poland. His major research interest is wheel–soil interaction analysis with a focus on experimental studies. One of the practical applications of research results by Dr. Pytka is an assessment of grass runway surface in terms of determining the airfield performance of airplanes. He has been dealing with this subject for 12 years, and this monograph is a summary of a significant step in developing research methods.

Author biography

Introduction 1

1.1 FLYING ON GRASS AIRFIELDS

An airplane may operate on grass airfields due to three following reasons:

- operating from a base airport with a grass runway,
- flying intentionally into a known or unknown unpaved airfield, or
- landing on an unimproved, opportune site due to engine or any other airplane system failure.

Flying from well-maintained grass runway has its unprecedent advantages. First, the airplane structure is dynamically less loaded at touchdown and during the ground roll. Tires last longer than on paved runways as well as passenger comfort can be higher [1–3]. Flying into known or unknown terrain, which is not an airfield or airstrip, is very popular in the United States, Canada, or Africa. The so-called "bush flying" or "off-field flying" may be a kind of adventure [4], but, in many cases, plays a significant role in connecting with remote, inaccessible areas. Examples are so-alled missionary flights, delivering food, clothing, and supplies to Christian missionaries at American Indian reservations. Incidents and accidents are very frequent, mainly due to not having enough piloting experience and unfamiliarity with the site. One typical scenario is that a pilot lands safely on a grass airstrip to give a visit but is unable to get airborne due to too short distance available for takeoff [5–7]. As for the last reason, a field, a pasture, or a meadow is used as an emergency landing site. In most situations, an airplane is partially damaged or even totally destroyed, but chances for a pilot and passengers to survive are relatively high. Two factors have a dominant role for safe and successful operation: (1) surface condition and (2) the length of the available runway and the surrounding obstacles.

Flying from grass airfields is the domain of small light sport, private, or cargo planes. However, in recent years, there has been a strong trend toward adapting high-performance airplanes to take off and land on grass or gravel/

DOI: 10.1201/9781003312765-1

dirt surfaces. An example is the Pilatus PC-24, which is so far the only business jet-certified to operate from unpaved airfields, including grass airfields.

1.2 GRASS AIRFIELDS AROUND THE WORLD

Historically, grass fields grown in the first airports and grassy airfields are still very popular. In the United States, there are about 11,000 grass fields which add to the potential of air transport system and improve safety since these fields may also be emergency landing sites.

Airfields with grassy runways are also very popular in Europe. The General Aviation Small Aerodrome Research Study (GASAR) analyzed 687 aerodromes in England which come under the scope of GA, classifying 374 into 6 types. These range in size from those of regional airports to that of the smallest farm strip, although 84% of GA flights operate from 134 of the larger aerodromes in the first four categories. The factors used in determining how an individual aerodrome is categorized by the GASAR study are based broadly on size and facilities. The six types of aerodrome are described, in size order, as regional airports, major GA airports, developed GA airfields, basic GA airfields, developed airstrips, and basic airstrips [8, 9].

A great number of grass airfields are located in Western Europe. In France, Germany, Switzerland, or Austria, a wide use of grass fields is related to the huge popularity of gliding/sailplane flying and air tourism. In sailplane cross-country flying, landings due to the lack of thermals occur quite frequently. Evacuation tow flights from opportune landing sites require that takeoff distance has to be calculated taking into account the drag of both towing airplane and sailplane. In Alpine region, at high altitude, airfields' density altitude is a significant factor having an effect on the performance of the unpressurized engines and, finally, the ground performance of the airplane. In Poland, only 56 of all 335 airports have paved runways, and the remaining are grassy airfields or airstrips. This is significant and prompts the serious consideration of grass airfields as an important element of air transport infrastructure. A significant number of those airfields are well maintained, with grass surface rolled in order to increase the bearing capacity and reduce surface roughness. They are operated by aeroclubs and provide seasonal tower control, aviation fuel, and weather service. Typical maintenance routines include mowing, rolling, and chemical treatments. Other grassy airfields are private airstrips of various conditions. A similar situation can be found in most of East-European countries.

One extremely important advantage of grass airfields is their harmony with nature. The grass runway is itself a local ecosystem, a habitat for micro- and macro-organisms, including insects, amphibians, reptiles, birds, and even small mammals. Despite a seemingly dangerous conflict of interest, a symbiosis usually occurs, and the fauna gets used to the conditions at the airport, while the people – users of the grass airport – accept the presence of the animal world. An example is the Świdnik grassland airport, in the area of which there is a colony of pearl gopher, a rare and protected rodent species. Of course, safety procedures are essential in case of potential collisions with intruders. Accidents for this reason are extremely rare, and we hear more often about collisions of a communication plane with birds in the area of typical airports with a paved runway.

1.3 FLYING FROM GRASS AIRFIELDS – MAJOR PROBLEMS

Grass airfields have worse mechanical parameters of the runway surface than surfaced airports. Most of the problems result from the deterioration of the mechanical strength of the surface caused by weather conditions. As a result, grass airfields stay closed during the winter months as surface conditions are uncertain. Moreover, incidents and accidents are caused by the lack of pilot experience and unfamiliarity with a given airfield [10–12]. It is also a very important factor that there are no official guidelines for the use of grass airports. The influence of soil and grass deformability and high sensitivity to weather conditions make it more difficult to predict the performance of the aircraft on a grass runway. A typical grass runway in dry conditions provides safe takeoff and landing conditions for aircraft up to 5,700 kg but may be too soft for light aircraft or sailplanes after heavy rain. Most importantly, however, there is no system similar to Notice to Airmen (NOTAM) runway friction measurement and reporting for paved aerodromes that can help determine the airplane performance on a grass runway.

1.4 PURPOSE AND SCOPE OF THE PRESENT WORK

The aim of this study is an attempt to systematize research methods, mainly experimental, for the analysis and modeling of the airfield performance of an aircraft operating from a grass runway. The author set himself the goal of developing new measurement methods as well as identifying (parameterizing) models describing the influence of significant parameters on airfield performance at a grass airfield. Chapter 2 contains an analysis of the current state of knowledge and technology in the discussed topic. Factors important to airfield performance are discussed. Design of airplanes that typically operate from a grass airfield is reviewed. Chapter 3 proposes an analytical model describing the performance of the landing gear wheel as well as the airfield performance of aircraft on a grass runway. The next chapter presents the methods and instrumentation used in the ground and flight tests.

The wheel dynamometer, the artificial-intelligence-based method for measuring the takeoff and landing distance, as well as the portable wheel–surface tester for determining the rolling resistance and braking friction coefficients are described in details. Chapter 5 contains the results of ground and flight test measurements and their analysis, while the last chapter of this book describes a project of an online system that provides information about the surface condition of a given grass airfield. A beneficial effect of the content presented in the monograph may be the practical improvement of flight safety on the grass runway, as well as the increase in the use of grass airfields within the existing air transport system and also in the future.

1.5 SUMMARY

This chapter presents the issue of grass airfields and their use in GA and air transportation system as well as the most important problems resulting from the fact that the performance of the aircraft on the grass runway is usually deteriorated than on a paved aerodrome. The advantages of grass airfields are presented, and the content of the following chapters of the monograph is briefly presented.

REFERENCES

1. Cook, L., "Getting High off Grass. Flying Organically with Turf Runways", *Kitplanes*, October, 2016, pp. 24–27
2. Hirschman, D., "No Runway? No Problem", *AOPA Pilot*, March, 2014, pp. 58–65.
3. O'Quinin, R., "Grass Landing", *Sport Aviation*, July, 2010, p. 76.

4. Mandes, G., "Alaska Wild Flying", *Flying Magazine*, June, 2013, pp. 56–61
5. Garrison, P., "Sunday Drive", *Flying Magazine*, August, 2013, pp. 40–42.
6. Krawcewicz, K., "Czy samolot wyjdzie?" (Will an Airplane Takeoff?), *PLAR Przegląd Lotniczy Aviation Revue*, 8/2011 (204), pp. 40–43
7. Landsberg, B., "Soft Field, Soft Thinking. Who Is Responsible for a Takeoff Gone Wrong?", *AOPA Pilot*, September, 2016, p. 20
8. Report on the Light Aviation Airports Study Group (LASSG), Civil Aviation Authorities, UK, 2005
9. www.lightaircraftassociation.co.uk/Consultation/laasgTheIssue.html [cited 15 February 2017]
10. Mauch, H., "Fit fuer den Acker", *Fliegermagazin*, März, 2012, pp. 58–61
11. Mauch, H. and Kutschke, E., "Tiefe Boden – hohe Kunst?", *Fliegermagazin*, Nr. 1, January, 2016, pp. 64–67
12. Schiff, B., "Legendary Popularity", *AOPA Pilot*, February, 2014, pp. 50–57.

Airfield performance of an airplane – the state-of the-art

2

2.1 TAKEOFF AND LANDING

A general schematic of the takeoff and landing mechanisms is depicted in Figure 2.1. Here, takeoff and landing are divided into phases, as well as points, namely 1 – starting point, 2 – rotation, 3 – liftoff (during takeoff), 4 – climb as well as 5 – approach, 6 – flare, 7 – touchdown, and 8 – stopping point. The diagram shows the ground roll distance for takeoff, $S_G^{T\text{-}O}$, and for landing, S_G^L.

In takeoff analysis, we have the following forces acting on an aircraft (see Figure 2.2): thrust T, aerodynamic drag D, weight W, aerodynamic lift L, and rolling resistance of the wheels F_{RR}. The aerodynamic forces are functions of aircraft velocity and change during the takeoff (they increase as the velocity increases). The thrust, which is generated by the propulsion unit (engine + propeller or turbine), is the force which acts against aerodynamic drag, rolling resistance, and an inertial force. The thrust also changes during the takeoff as a function of velocity (decreases to approximately 70% of the initial value). At the start point, the weight is the only force in vertical direction, but during ground roll, the lift force is generated and this force decreases wheel loads. When the lift equals the weight, the aircraft flies and the thrust is consumed by the aerodynamic drag, climbing or accelerating the aircraft. The acceleration

DOI: 10.1201/9781003312765-2

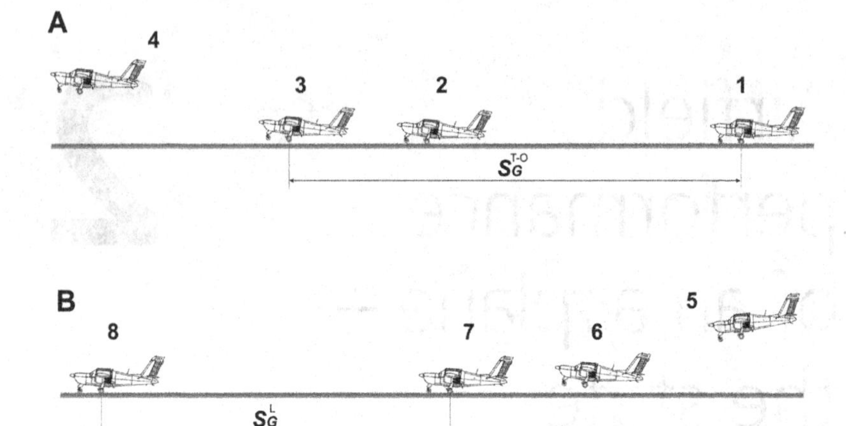

FIGURE 2.1 A schematic of takeoff and landing mechanisms.

of the aircraft during takeoff ground roll can be expressed by the following equation [1]:

$$a=\frac{g}{W}\left[T-D-k_{RR}\left(W-L\right)\right]=g\left[\left(\frac{T}{W}-k_{RR}\right)+\frac{\rho}{2\left(\frac{W}{S}\right)}\left(-C_{D0}-KC_L^2+k_{RR}C_L\right)V^2\right] \tag{2.1}$$

where: C_{D0} – drag coefficient, K – ground effect coefficient, C_L – lift coefficient, and V – ground velocity of the airplane.

The ground roll distance is determined by integrating velocity divided by acceleration:

$$S_G=\int_{V1}^{V2}\frac{V}{a}dV=\frac{1}{2g}\int_{V1}^{V2}\frac{d\left(V^2\right)}{K_T+K_AV^2}=\left(\frac{1}{2gK_A}\right)\ln\left(\frac{K_T+K_AV_f^2}{K_T+K_AV_i^2}\right) \tag{2.2}$$

with

$$K_T=\left(\frac{T}{W}\right)-k_{RR} \tag{2.3}$$

$$K_A=\frac{\rho}{2\left(\frac{W}{S}\right)}\left(-C_{D0}-KC_L^2+k_{RR}C_L\right) \tag{2.4}$$

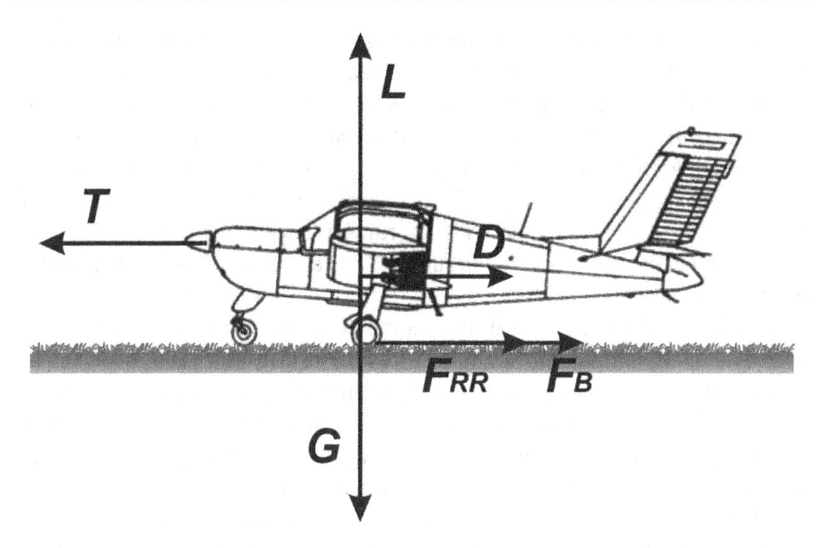

FIGURE 2.2 Forces acting on an airplane during takeoff run.

Landing is much like taking off – but in reverse. Eq. (2.2) can be used to calculate ground roll during landing with respect to change the terms of velocity. The thrust force during landing is called the idle thrust for conventional aircraft. When an aircraft is equipped with a thrust reverser or reversible propellers, the resulting thrust will be negative. Also, the aerodynamic drag may be increased by spoilers, speed brakes, or drag chutes. There are often obstacles around the airfield, so it is practical to determine not only the ground roll, but also the total takeoff distance from the start to the point where the aircraft is 15 m above the ground. Note that the aforementioned method assumes the constant value of rolling resistance coefficient, k_{RR}.

2.1.1 Takeoff and landing distance measurement

Measurement of the takeoff and landing distance is one of the mandatory procedures in the certification of the aircraft and is of great practical importance. Knowledge of these performance is a prerequisite for safe flight performance, especially when flying at airports with an unpaved runway. Various methods of measuring takeoff and landing are used in industrial and research practice. Given next is a brief overview of selected methods.

Sighting bar method uses one or two sighting bars located at a known distance from the runway. By the use of these bars, a stopwatch, runway observers, and hand-recorded flight data, data of takeoff and landing distances may be obtained. In the triangulation method, a camera and a scale located near the camera, parallel to the airfield direction, are used. The scale has two wires installed so that the gap between the scale and the wires ("runway wire" and "50 ft wire") is set to coincide with the runway center line and the 50 ft (or 15 m) screen at the mid-point of the runway. The device is set level and parallel with the runway.

Movie theodolite method utilizes a movie camera that is used to record other data such as time and azimuth in addition to filming the airplane. In onboard theodolite method, a camera is mounted on the airplane to obtain three-axis position information. Runway lights and other objects on or along the runway of known size are used with photogrammetric techniques and perspective geometry to obtain airplane position and altitude.

Methods based on electronics include Del Notre Transponder, laser altimeter, and GPS (Global Positioning System) sensor. Del Notre Transponder measures horizontal distance and, when combined with a radio altimeter for height information, provides all the distance and height information necessary for determining takeoff and landing distances. The Transponder consists of a Distance Measuring Unit (DMU); a master transponder; and associated antennae, cables, and power sources.

The laser method uses a laser landing altimeter. This unit measures distance using a modulated laser beam with centimeter accuracy.

A GPS sensor is used to determine aircraft longitudinal and vertical position together with time coordinates during a takeoff or landing. The weakness of this method is that the moment of liftoff is difficult to be determined with required accuracy.

2.1.2 Takeoff and landing measurement with the use of GPS and IMU

Pytka et al. (2006) have proposed a method in which the point of liftoff or touchdown of the airplane is recognized from time courses of vertical acceleration of the airframe, captured during takeoff and landing. Then, knowing the starting point and rollout end point, it is possible to determine the takeoff or landing distance [2].

The moment of touchdown can be recognized on the basis of measurements made with the use of an accelerometer located at any point in the aircraft structure. The acceleration pulse is characterized by very high dynamics;

moreover, the observed acceleration is multidirectional. The accelerometer can be placed on the floor in the cabin of the aircraft. It is clearly visible at the moment when the acceleration amplitude increases rapidly and the area with the highest amplitude disappears after a time (the airplane breaks off the ground for a moment, jumps up) to then reappear and continue most of the rollout distance. Likewise, it is possible to recognize the remaining phases of an airplane landing [3].

In conclusion, it should be noted that one important weakness of this method was that the points of start, liftoff, touchdown, and stop were determined manually in a post processing analysis. This excludes the method from a practical use in flight tests. However, the cited study has become the basis for the development of an improved method using artificial neural networks, which is detailed in Chapter 4.

2.1.3 Liftoff

The moment at which the airplane leaves the runway is known as the liftoff. At this moment, the contact between the tires of the main landing gear wheels and the surface of the runway is broken. In very rare cases, the front wheel may tear off last, as in, for example, de Havilland Canada DHC-4 Carribou with its famous stunt maneuver "wheelbarrow". Mechanically, the lift balances the plane's gravity, followed by a phase of flight with increasing speed and height gain. After reaching a certain horizontal speed, usually of 10–20% greater than liftoff speed, the pilot enters the plane into the phase of steep climb with parameters strictly defined in the flight manual.

Liftoff is usually preceded by the moment of the rotation, which consists in the fact that during the run-up, after reaching a certain speed (= rotation speed), the pilot slightly raises the nose of the plane so that the configuration of the plane corresponds to the angle of attack at which the lift reaches a safe maximum.

2.1.4 Airplane stability during takeoff and landing ground roll

Small- and medium-sized general aviation (GA) airplanes usually have a three-point landing gear, which consists of two main wheels and a front or rear (tail) wheel. In the first case, we are talking about a chassis with a front wheel, while in the second case, we have a so-called classic landing gear, with a tail wheel. The main landing gears usually have single road wheels. The front or the tail wheel is usually smaller than the main landing gear wheels due to lower loads it has to accommodate. For large transport airplanes, landing gears may be similar, that is, three-point, but the number of wheels per one gear is usually greater than 1, and for the main landing gear, it may be even greater than 10. In large transport aircraft, other than three-point landing gear, systems are also used.

The three-wheeled running gear is characterized by specific mechanics in terms of maintaining the direction of movement and driving stability. The location of the center of gravity between the main supports and the front or rear (in the side projection) guarantees stability around the transverse axis. Thanks to this, the plane does not tend to fall on its tail or roll over during taxiing. The exception are airplanes with a tail wheel with a high center of gravity, for example, the PZL 104 Wilga, which in this respect is quite difficult to pilot and requires attention during touchdown. Lateral stability is important due to the possibility of turning the plane around the vertical axis and consequently changing the direction of movement until veering off the runway. Often, landings take place in crosswind, which is the most important factor disrupting the aircraft's movement.

Weick (1934) in Abzug (1999) [4] conducted an experiment with simple models of a three-wheeled undercarriage in two versions:

- a system with a front wheel suspended on a rotating fork as a castering wheel and
- rear (tail) wheel arrangement.

The geometry of the tested models of running gear was the same. Weick's experiment consisted in running both models on an inclined plane and observing the behaviors of both models. It turned out that the system with the front trailing wheel always maintained the stability of movement, even if it was extended with a certain initial angle of deviation of the longitudinal axis from the direction. In contrast, the system with the rear wheel always rotated around the vertical axis, so its movement was directionally unstable.

Abzug (1999) conducted analytical studies of the stability of Weick models, using a model with six degrees of freedom. The perturbation theory was applied, used to find an approximate solution to a problem that cannot be solved in a strict manner due to the degree of compilation or the variability of significant parameters, a small parameter that determines the deviation (disturbance) from the accurately solved problem. The first segment of the series is a solution to a solvable (undisturbed) problem, while the subsequent members of the series describe a deviation from a strictly solvable problem.

Abzug (1999) used the perturbation theory to analyze the longitudinal stability of an airplane with a landing gear for both cases of front wheel and tail wheel, for a glider with an unsprung single-wheel landing gear and for the model Weick system. The assumptions were as follows [4]:

- ground roll without lateral movement, which gives the tested system two degrees of freedom (lateral drift and yaw movement);
- weight of wheels and legs reduced to the weight of the airplane; and
- a multi-wheel system replaced by a single wheel system with a replacement wheel.

The basic information for the analysis of the stability of the aircraft rollout in contact with the ground is the tire characteristics. When the plane touches the ground (touchdown point), the aerodynamic forces are dominant, but during the rollout, when the speed decreases, the forces in the wheel–surface system become so large that their values significantly affect the longitudinal dynamics of the plane. Zero longitudinal slip was assumed. The characteristics of unit forces and moments on the running gear wheels were adopted from the literature, and in the case under consideration, these were the characteristics of the tire–asphalt surface system, obtained in experiments on drum stands.

After establishing the dependence on the forces acting on the airframe and applying the assumption of small disturbances, the derivatives of forces and moments on the wheels were determined in relation to the sideslip angle and the yaw rate. From a comparison of these derivatives for two cases:

- landing gear wheels on rocker arms and
- rigidly suspended wheels,

it is shown that the type of suspension has a significant impact on the stability of the tested undercarriage.

The main study of the stability was based on the observation of the location of the roots of the characteristic equation. It has been shown that there is an instability (true component greater than 1) when the main landing gear is in front of the center of gravity.

The method adopted by Abzug allows to study the longitudinal (directional) stability of the aircraft at different speeds and for different positions of the main landing gear. It does not take into account the effect of deformable ground, which significantly affects the characteristics of lateral forces in the wheel–ground system. Deformability of grass surfaces is mainly due to the presence of above-ground vegetation. In Chapters 4 and 5, an experimental method together with results on aircraft stability during rollout on grass runway is presented.

2.1.5 Touchdown

A critical moment in the landing is the touchdown. At that time, temporary contact forces act between the surface and the wheel. Both the surface and the wheel deform significantly. A great amount of an airplane's kinetic energy is dissipated into soil deformation energy, shock absorber energy, and tire deflection energy.

The most important forces acting on the airplane's wheel during a touchdown are as follows [5]:

- total load force, F_Q, which is a result of aerodynamic lift forces, aircraft weight, Q, and inertial forces during the time period of the touchdown and
- soil reaction, R.

The total load force, F_Q, is a sum of elementary forces acting along the total contact area between the tire and soil surface. This force is distributed along the total contact area, S, and can be determined as an integral:

$$F_Q = \int F_Q^{\alpha} d\alpha + \int F_Q^{\beta} d\beta \tag{2.5}$$

The load force distribution is a function of tire deflection, contact area surface, and soil mechanical strength. The load components are functions of α and β, angles of wheel–surface interaction area:

$$F_Q^{\alpha} = F_Q\left(\frac{1}{\sin(\alpha)} + \cos(\alpha)\right) \tag{2.6}$$

$$F_Q^{\beta} = F_Q\left(\frac{1}{\sin(\beta)} + \cos(\beta)\right) \tag{2.7}$$

Aerodynamic forces – lift, drag, and propeller thrust – are functions of air speed of the airplane. The lift force acts vertically and equals the airplane's weight at liftoff. This force reduces the vertical load on the undercarriage wheels. These three forces change their values during landing and takeoff. Lift decreases as an effect of airplane's deceleration. In effect, vertical loads on the landing surface increase and reach a maximum of Q at $V = 0$. In the moment of touchdown, the aircraft weight is balanced by lift force, so the wheel loads are mainly the temporary inertial forces:

$$F_{AV} = Q\frac{a_V}{g} \tag{2.8}$$

$$F_{AH} = Q\frac{a_H}{g} \tag{2.9}$$

Ground reaction is important as a basic input parameter in the load–stress analysis of an aircraft's undercarriage. Forces acting on a soft surface cause deflections; therefore, the vertical loading force is lower. In the moment of touchdown, ground reaction has both vertical and longitudinal components (see Figure 2.2). In a short period after a touchdown, the aerodynamic forces are constant as the velocity does not change significantly, and, therefore, the vertical reaction on aircraft wheel can be expressed as shown here:

$$R_V = b\int \sigma \sin(\varphi)d\varphi \tag{2.10}$$

and the horizontal reaction as:

$$R_H = b\int \sigma r \sin(\varphi)d\varphi \tag{2.11}$$

where σ = stress on a tire–ground contact surface and b = tire width.

In Chapters 4 and 5, a study on experimental determination of soil reaction in term of soil stress has been included.

2.2 FACTORS HAVING AN EFFECT ON AIRFIELD PERFORMANCE

2.2.1 Effects of airplane design

A great number of today's GA airplanes are suitable for grassy airfield operations. Certainly, there are types which have been especially designed to be used on unsurfaced landing sites. Generally, STOL airplanes are best suitable for off-field flying. Some important elements of airplane design are discussed next.

2.2.1.1 Powerplant

A simple way to obtain good airfield performance is by means of having high power-to-weight ratio, which can be achieved by using of powerful engines or lightening the entire design. The latter is especially popular in small, light sport or homebuilt airplanes. The use of transmission gear like on Rotax 912 engines or AI-62R radial engine on the PZL 104 Wilga allows to use low RPM propellers of greater diameter. This allows to increase thrust, maximize propeller's efficiency, and decrease noise. For heavier crafts, turbine power is the best option, also because this type of engine enables reverse thrust during landing.

One of the possibilities of improving the airfield performance is the modification of the propulsion system of the serial aircraft. An example is the Boss 182 aircraft, which was based on the Cessna 182 aircraft. The original 230 HP IO-540 Lycoming engine was replaced with the Lycoming IO-580 engine with 315 HP. As a result of this change, the takeoff distance was shortened from 243 m to 195 m, and the takeoff distance of 50 ft was shortened from 462 m to 394 m.

The radical improvement in ground performance was possible thanks to the replacement of the piston engine with a turbine engine, as in the case of the Wilga DRACO aircraft, described in the following section.

2.2.1.2 Wing

Basically, wing design including its profile and planform determines lift force characteristics. For short takeoff distance, the more lift generated on the wing the better, but this should be achieved at a possible low airspeed. Turbulent airfoils of about 12–16% thickness are frequently used together with rectangular wing planform.

Wing design elements such as a high-lift wing section, low wing loading, wing mechanization (Fowler flaps, flaperons, slats), vortex generators, and leading edge cuffs have a positive influence on short and soft field performance of the entire airplane [3, 6]. Those features decrease stall or minimal speed and improve stability and maneuverability at high angle of attack. One drawback of the wing mechanization is its complexity, while the use of high-lift profiles results in lower cruise speed.

Perhaps the most efficient way of shortening the takeoff roll is the use of flaps. This is because the SG is very dependent upon the takeoff speed. However, the shortest ground roll does not always mean the shortest total distance over an obstacle, and important factors that affect air distance phase include wind, aircraft weight, air density and temperature, ground effect, and of course, piloting technique.

2.2.1.3 Landing gear

Landing gear wheels with high diameter, so-called "tundra" tires, ensure the improvement of field performance in the sense of better accommodation to surface roughness and additional damping by tires. Moreover, the A-frame landing gear, together with shock absorbers or telescopic springs (like in the WWII Fieseler Storch), helps a lot during landing with high descent rate. One interesting solution is the use of rocker-arm-type suspension with air absorbers, like on the PZL 104 Wilga 35A airplane. It works efficiently on a rough terrain, without the use of big "tundra" wheels, which adds pronounced to aerodynamic drag. Wheel brakes of high efficiency are necessary for shortening the landing ground roll.

2.2.2 A brief review of airplanes

Basically, a typical STOL airplane is two-to-six seater, single engine, with high wing machine. One with the best reputation is the Cessna 180/185 family, the utility airplanes with great soft and short field capabilities. Although not in production since 1985, the airplanes are still in the use all over the world. A high wing, tailwheel design, powered by 230 to 260 HP direct drive engines with a strong and light airframe is an example of a classic. Similar airplanes are PLZ 104 Wilga, Helio Cruiser, and Aviat Husky. In bigger, 10–15 seat segment, we find the Antonov An-2, Cessna Caravan, Pilatus Porter, Quest Kodiak, and the twin-engine Cessna Skycourier. These airplanes are powered by big/high displacement piston or turbine engines, which ensures enough thrust for short takeoff ground roll. In contrary to all other designs, the Caravan, Kodiak, and the SkyCourier have nose wheel landing gear. There is a grooving segment of business-class high-performance airplanes that, although not typical STOL, can operate from unpaved, grass, or gravel airfields. This includes Swiss-made PC-12 turboprop- and PC-24 turbofan-powered aircraft, which is in fact the first certified jet capable of grass field operation.

The aforementioned DHC-4 Caribou plane is also an excellent STOL plane. It is perfect for operation in difficult, unprepared runways. But there are more interesting military airplanes with STOL capabilities, mainly logistic/cargo planes (Antonov An-26, An-72, C-17A Globemaster, C-4 Hercules). These airplanes are heavy; so are the loads exerted on the surface, therefore, their landing gear design utilizes multiple-wheel systems.

Perhaps the most numerous segment create small and light LSA (Light Sport Airplane) or homebuilt airplanes. Typically, with tube and fabric structure design, powered by Rotax or small Continental and Lycoming engines, these crafts accommodate two persons and allow to operate from unimproved airfields or even in "no-runway" conditions. Here, we have the classic Piper Cub and its clones (Cub Crafters, Legend Cub, Wag Aero Sport Trainer), ACA Decathlon, Zenair STOL with all metal construction, Criquet Aviation Storch replica, the Bearhawk, and the latest design of Just Aircraft, the SuperSTOL. Among some unorthodox projects, we'll find the Doubleender, an interesting design with typical long-stroke landing gears and bush wheels and surprisingly unconventional power by two in-line engines. It is to point out, however, these aircraft are especially designed to operate from poor, unsurfaced airfields, while the main goal of this study is to analyze the ground performance of typical GA airplanes and to determine their limits with respect to surface conditions.

2.2.2.1 The Wilga DRACO

One interesting design is the Wilga DRACO airplane (Figure 2.3), which was created as a conversion of the serial PZL 104 Wilga 2000MA aircraft. The constructor, Mike Patey, used many modifications to improve flight and ground performance, and the design assumptions in this regard were as follows:

- ultra-short takeoff and landing;
- steep approach to landing, with a low approach speed;
- possibility of taking off and landing on unprepared, uneven, and unpaved terrain.

FIGURE 2.3 The Wilga DRACO SuperSTOL airplane.

Photo courtesy: Mike Patey

Patey achieved these intended ground performance through a series of modifications, including:

- change of the aircraft propulsion system, the Lycoming IO-540 internal combustion engine with a capacity of 300 HP was replaced with a Pratt & Whitney PT6A-28 turboprop engine of 680 HP;
- increasing the wing's area by widening the chord (a nose part was riveted to the original wing, thus modifying the wing profile), new wing tips as well as increasing the surface of the vertical and horizontal tail;
- modification of the fuel system and the installation of fuel tanks in new, enlarged landing gear fairings;
- landing gear conversion: new wheel axles, brakes, large low-pressure tires, new enlarged tail wheel;
- modifications to on-board installations, including lighting, oxygen installation; and
- new instrument panel with glass cockpit avionics and new crew/passenger seats.

Compared to the serial plane, Wilga DRACO has much better performance and is characterized by improved handling qualities. Cruising speed and climb speed increased while stall speed decreased. The range of Wilga DRACO is comparable to the base plane, but the maximum altitude is significantly higher. The ground performance of the aircraft after the conversion is also more favorable: the takeoff distance of Wilga DRACO is 41 m, and the landing rollout distance is 49 m (the values for the serial aircraft are 350 m and 270 m, respectively). Table 2.1 includes a comparison of selected design parameters and performance of both airplanes.

TABLE 2.1 Performance comparison PZL Wilga versus Wilga DRACO

PARAMETER	*PZL 104MA WILGA 2000*	*WILGA DRACO*
Takeoff mass, kg	1,500	1,815
Engine power, HP	300	680
Rate of climb, m/s	4.7	15.24
Cruise speed, km/h	208	300
Stall speed, km/h	93	61
Takeoff distance, m	350	41
Landing roll, m	270	49
Altitude, m	3,500	8,534
Range, km	1,240	1,287

2.2.2.2 The Scrappy airplane

The Scrappy airplane is another design by Mike Patey. The Scrappy plane is based on the Carbon Cub EX-3 bush airplane. It is powered by a custom engine, which was created as a modernization of the Lycoming 720 engine. Serial engine modifications included porting, polishing, high-compression modifications, a scratch-built induction system, custom head lockers, and an electronic ignition conversion. The engine power is 500 HP, and a short-term increase of a further 250 HP is possible as a result of the nitromethane addition.

An interesting innovation was the use of a test concept propeller using air boat blades designed to pull without stalling. The main goal was to obtain optimal thrust during takeoff. It is a propeller with a construction typical for an airboat, where considerable thrust is required at low speeds (see Figure 2.4). Unlike the serial propeller designed for the Carbon Cub airplanes, the propeller used by Patey does not cavitate. It is a four-blade, fixed pitch, ground adjustable propeller.

The custom wings include double slats on the leading edge and large flaps and drooping ailerons on the trailing edge for slow-speed handling. Without the slats, almost full elevator authority would be required to counteract the flaps and drooping ailerons combined to keep the nose of the aircraft up. When the slats are deployed, the wing chord enlarges by 14 inches, which in turn allows reshaping the camber line to yield a greater lift. As a result, the

FIGURE 2.4 The propeller used by Mike Patey in his Scrappy airplane design.

Photo courtesy: Mike Patey

FIGURE 2.5 A schematic and the view of the wing mechanization in the Scrappy airplane.

Photo courtesy: Mike Patey

wing has a larger area, which is important when flying at low speed. In addition, the center of pressure shifts to a location that allows full forward and aft range of the stick and less trim. The slats, flaps, and drooping ailerons are connected, and they all can be deployed from a single trigger in order to simplify wing transformation. The profile of the wing with slats and flaps is shown in Figure 2.5.

An interesting innovation in the Scrappy aircraft is the wheels' suspension that includes double shocks and an airlift that can lower the front end and lift the tail (Figure 2.6). This was primarily to improve the visibility during taxiing. In addition, the suspension is articulated on the left and right sides, which is important when landing or parking on slopes with an angle. The Scrappy chassis is fully dynamic and adjustable during the flight, allowing to change depending on the changing load (e.g., due to fuel consumption). The suspension

FIGURE 2.6 Landing gear of the Scrappy allows to adjust the airplane position.

Photo courtesy: Mike Patey

can also be changed on the basis of terrain, weight, and wind to optimize landing. The position (adjustment) of the wheels is constantly measured by means of potentiometric sensors; in addition, the pressure in the shock absorbers is also monitored. Measurement data is analyzed and correlated with the load and a computer system enables actuation to be achieved in the selected mode [6].

2.2.3 Airplane certification tests for unpaved runway operation: the Pilatus PC-24

Type certification for unpaved runway operation requires to conduct a test campaign in order to show any effect of increased roughness on the structure and systems of the aircraft, including landing gears and their attachments. Aircraft ground loads appropriate to the surfaces to be approved for operation shall be considered in meeting the appropriate certification requirements. Fatigue and damage tolerance, including accidental damage inspections and life limits for landing gear and other structural elements, should be reviewed and revised as appropriate. For taxi, takeoff, and landing roll, longitudinal runway profiles on which the airplane will be operating should be directly measured. Then, the airplane response using those profiles should be analyzed. Alternatively, measuring the airplane response during a ground test on those runways should be performed. Fatigue and vibration effects should also be evaluated. In these cases, aircraft mission profile should be assumed in term of percentage of operation on rough or unpaved runways.

The Pilatus PC-24 aircraft is so far the only business jet certified to operate from unpaved airfields, including grass airfields.

It is interesting to note how the certification tests of this aircraft were carried out in order to be approved for use at unpaved airports. The test program along with the procedures and measuring and test equipment used are briefly described later in the chapter.

Ground performance testing at airports with unpaved runway surfaces continued in parallel with the final type certification process. On the disused runway of the Buochs test airport in Switzerland, chalk tests were carried out on the pavement. By depositing on the airframe, the chalk made it possible to determine the effects of contamination on the underside of the aircraft, tires, structure, and engines. Based on the results obtained, a detailed research plan was prepared covering approximately 2 years of testing on five surfaces and three uncured substrates. These were the following surfaces: grass, sparse vegetation, sand, gravel, and compacted gravel of the Arctic type. The research was started on the grassy runway in Sion, the runway of which is 2000 ft long, so no flights were made there but only the measurements of shock loads on the shock absorbers. The results of these measurements were the reference values to which the results of similar measurements performed on each of the tested surfaces of the runway were compared.

An important aspect of the test program was runways with representative surface characteristics. The tests on the gravel surface resulted from the need to adapt the aircraft to the requirements of the Royal Flying Doctor Service (RFDS) in Australia, from which an order for multiple aircraft was received. The PC-24 order from the Royal Flying Doctor Service (RFDS) in Australia was of particular importance to Pilatus as RFDS wanted to be able to work off-road as a standard operational requirement. The search in Europe for an area similar to the typical unpaved airfields in Australia was focused on Spain due to soil and climate conditions. RAF Woodbridge airfield in Suffolk was chosen with its 1,700-m long gravel/dirt runway (Figure 2.7). The A400M had previously been tested on the Woodbridge runway.

A few weeks before the start of the actual trials at Woodbridge airport, the ground crew prepared technical support, including earthmoving and agricultural machinery. Fire trucks with crew; three containers for storage, check-in, and domestic use; water supply; and electricity generators were also rented. The minimum team at any given time consisted of about 12 people. It was also necessary to provide meals for team members working in the field.

FIGURE 2.7 Tests on the gravel/dirt runway of the RAF Woodbridge airfield in Suffolk, UK.

Photo courtesy: Pilatus

Flight test engineers used rugged laptops placed in dedicated onboard stations. The laptops were linked to Flight Test Information (FTI). The FTI collected data on approximately 6,000 aircraft parameters and was linked to the Data Acquisition System (DAS). Both computer systems were supervised by an engineer who could observe selected data in real time. The first and second PC-24 prototypes were equipped with the FTI that weighs approximately 1,000 kg. The measuring equipment was housed in the housing, on the lower right side of the cabin, with the FTE installations on the left side.

The flight test campaign was driven by the weather to some extent, particularly the wind. In the case of performance and handling testing, the maximum wind speed was 5 kts. Therefore, most of the test flights were carried out in the morning and evening hours. Similarly, it was important to ensure the assumed soil conditions, especially humidity. Equivalent soil moisture corresponded to a rainfall of 4 l/m^2 (liters of water per square meter). This required the use of a 24 m farm sprayer and an expenditure of 170,000 l/day (liters of water per day). The numbers show the degree of difficulty of the implementation of flight tests with the use of real objects in real conditions.

FIGURE 2.8 Tests on the wet grass surface, Kunovice airfield, the Czech Republic.

Photo courtesy: Pilatus

As required by the European Aviation Safety Agency (EASA), tests to obtain the certification for grass runway operations must be performed on a variety of surfaces. For the tests of the Pilatus PC-24 aircraft, the trailing runway airports in Kunovice in the Czech Republic, Poitiers in France, and Duxford and Goodwood in Great Britain were selected. Dry grass tests were carried out at the Goodwood airport. The Goodwood runway surface was in a very good condition, thanks to previous maintenance and repairs. The airport in Kunovice has two long grass runways, a hardened runway as well as an extensive infrastructure necessary for aircraft tests. A detailed suite of tests was carried out there, including single engine takeoffs, engine failure and loss of spoilers, loss of the anti-skid system, etc. Subsequently, the wet grass taxiing tests were carried out, which was the beginning of the most difficult test, namely wet grass takeoff and landing, which was carried out after heavy rainfall. The results of all trials in Kunovice were very good (Figure 2.8).

One of the important parameters distinguishing the condition of the unpaved runway surface is hardness. The standard method for determining the hardness of a runway surface is by measuring with a cone penetrometer (see Chapter 4 for details). The penetrometric measurement results are then converted into the California Bearing Ratio (CBR), which is an universal measure of the hardness of various surfaces, ranging from concrete (CBR = 100); to gravel and dirt (CBR = 30); to loose soil (CBR = 7). Tests for the runway with sparse vegetation were carried out at CBR = 20 to 30. The hardness of the wet grass surface on which the tests in Kunovice were carried out had a CBR value of 2. Under these conditions, the ruts in soft soil, caused by aircraft tires inflated to 76 psi pressure, were 150 mm deep [7].

2.3 EFFECT OF AIRFIELD CONDITIONS

2.3.1 Definitions

According to the Certification Authorities for Large Transport Aircraft [8], in the sense of surface preparation, we can distinguish the following types of runways and taxiways:

- paved – surfaced with asphalt or concrete;
- rough – runway (paved or unpaved) with bump height to length curve above the "tolerable limit" curve of the ICAO Annex 14 Runway surface evenness criteria or worse than the similar runway criteria specified in AC/AMC 25.491;
- unpaved – a surface composed of unbound or natural materials. Unpaved surfaces may include gravel, coral, sand, clay, hard packed soil mixtures, grass, turf, or sod;
- semi-prepared – an existing runway or taxiway that may have required a considerable construction effort. Variable factors influencing roughness intensity can be foundation soil structure, climate (e.g., frost heave), size of aircraft, pavement composition, maintenance standards, and the number of aircraft using the runway;
- unprepared – an unsurfaced natural ground area typically suitable only for the operation of military cargo-type aircraft with little or no preparation [8].

Runways defined as semi-prepared best fit the topic of this book. For commercial applications, most gravel runways and some grass runways would fit into this definition. This type of field may be surfaced with landing mat or a protective membrane. Most grass airfields carry out a variety of leveling measures or improving bearing strength. Maintaining the grass airport and ensuring safety also requires the removal of foreign objects, more if the grass surface is a living ecosystem inhabited by animals, birds, etc.

2.3.2 Tire–runway interaction

During takeoff or landing, the interaction between aircraft wheel tires and the surface of the runway is the most important factor influencing airfield performance. Tire–runway interaction is characterized by means of two factors:

- coefficient of tire–runway friction, μ and
- rolling resistance coefficient, k_{RR}.

2.3.2.1 *Tire–runway friction coefficient*

Physically, the term μ comes from Amontons–Coulomb friction law and can be used to determine traction force, which, in the case of airplane during landing rollout, is equal the braking force F_B:

$$F_B = \mu m g \tag{2.12}$$

where m – airplane mass, g – Earth's gravity coefficient (= 9.81 m/s²).

This traction force does not act during takeoff, since the aircraft wheels are nor powered, with a exception of the so-called "electric taxiing" systems. In such systems, electric motors installed in hubs drive the wheels during taxiing, which reduces fuel consumption and noise in the close proximity of airport buildings.

During landing rollout, wheel brakes are used to shorten the distance, and the braking force is dependent upon tire–surface coefficient. Values of μ range from 0.2 for a typical tire on snow-covered surfaces; 0.3–0.5 for a tire on a typical grass runway; to 0.8 for a tire on a concrete or asphalt runway.

2.3.2.2 *Methods for tire–runway friction measurements*

Measurements of braking friction is a must for paved runways of airports in northern locations because of possible reduction of friction due to ice, snow, or water residues on runway surface. A number of different measuring devices have been developed for this purpose, and one of them, namely a friction tester, became definitely the most frequently used in the praxis. A typical wheel tester is a vehicle, typically a passenger car. As an example, in the SAAB Friction Tester, a measuring wheel with a 4.00 × 8′ tire was used in the measuring wheel with a tread and rubber compound composition similar to those used in aviation. The measuring wheel could be hydraulically lowered down to contact the surface and was set to rotate with a constant percentage of slip

(10% to 30%) relative to the car's constant speed (40 to 60 mph). Chain drive resistance was measured by a torque sensor, which then relayed this information to the on-board computer, which sent it to air traffic control in real time. The water tank was used to simulate wet conditions by means of an adjustable, constant flow of water in front of the measuring wheel [8].

The modernized SARSYS-ASFT measuring system uses the same general measuring principle (skiddometer principle). However, it has been modernized over the years, and now the system is installed on vehicles such as Volvo, Skoda, and VW. The measuring wheel is only connected to the rear axle by a ribbed spring attached to the pressing hydraulic cylinder of the wheel. This ensures a constant pressure (amounting to 1,400 N) and eliminates disturbances caused by vehicle movements (vibrations), which affect the repeatability of the measurement results. The compound composition, shape, and tread of the measuring tire are similar to those used in aviation or car tires, depending on the area of application. This ensures that the results are correlated with the actual friction values obtained during the operation of airplanes or cars. The rear axle, thanks to the use of a linear actuator and gas spring, allows you to start the measurement while driving. "Flying start" can be used at any speed without having to stop, which saves time and does not interfere with road traffic. Additionally, the built-in differential allows for measurements on bends, and the self-damping system in simulating wet conditions. The main computer is equipped with a touch screen to facilitate the function of the operator. It connects wirelessly to a measuring computer that controls the movement of the measuring wheel, the operation of the water pump, and the water pressure. The software allows for integration with optional equipment such as a GPS system or an ambient temperature sensor. It is possible to view data in real time, which are then archived [8].

A different principle is the use of a dynamometric trailer, which is a towed, single-axle vehicle with a measuring instrument and a skid system. One example is the MK2 GripTester, which is a dynamometric trailer that measures the braked wheel with a constant 15% slip, and the load and resistance are measured continuously. Due to its design, the device can be towed by most vehicles. It is characterized by low weight, little tow force, and a low center of gravity. This ensures stable and safe operation at any time of the year and in extreme weather conditions. The range of measuring speeds is from 5 to 130 km/h. The automatic water supply system ensures that the appropriate water flow is maintained for the accuracy of results at measuring speeds from 20 km/h to 95 km/h [8].

Another method is the use of a portable tester, which can be either a wheel tester, drag sleed, or pendulum devices. They are typically used in road construction industry; however, some concepts are of high potential for runway measurements.

2.3.2.3 Rolling resistance coefficient

The second important parameter describing the contact between the road wheel and the surface is the rolling resistance coefficient, k_{RR}. The rolling resistance coefficient synthetically captures the phenomena occurring in the area of interaction of the tire with the road surface, in particular those that counteract the motion of the road wheel. One of the physical interpretations explains the rolling resistance as the development of a moment of force acting against the rotation of the wheel. This moment is the result of tire and pavement deflection and is accounted for by moving forward from the point of the normal reaction force, as shown in Figure 2.9.

When a wheel is rolling over a hard (non-deformable) surface, the following components of rolling resistance F_{RR} are present:

- braking friction;
- inertia forces in case of unsteady motion (acceleration); and
- rolling resistance caused by local deformation of a tire and wheel slip.

Rolling resistance is physically related to M_Y, which counteracts the rotation of the wheel, as shown in Figure 2.9. The mathematical description is as follows:

$$M_Y = eQ \tag{2.13}$$

This consideration, although very close to real conditions, is difficult to use for calculations, mainly due to problems with determining the term e. Therefore, a proportionality coefficient k_{RR} was introduced, and, thus, we obtain:

$$F_{RR} = k_{RR} Q \tag{2.14}$$

where Q = vertical load of the wheel.

In the literature, sometimes, the rolling resistance coefficient k_{RR} is introduced as the friction coefficient, μ. In fact, this is not true, since friction cannot be identified with rolling resistance. We can speak of braking friction during breaking or traction during driving, when a horizontal force is generated on a tire–surface contact patch, and this force is physically similar to friction. By rolling, friction exists in wheel suspension (bearings). On the other hand, the major components of rolling resistance are tire and surface deformations, and therefore we use the rolling resistance coefficient, k_{RR}, which represents all components of rolling drag.

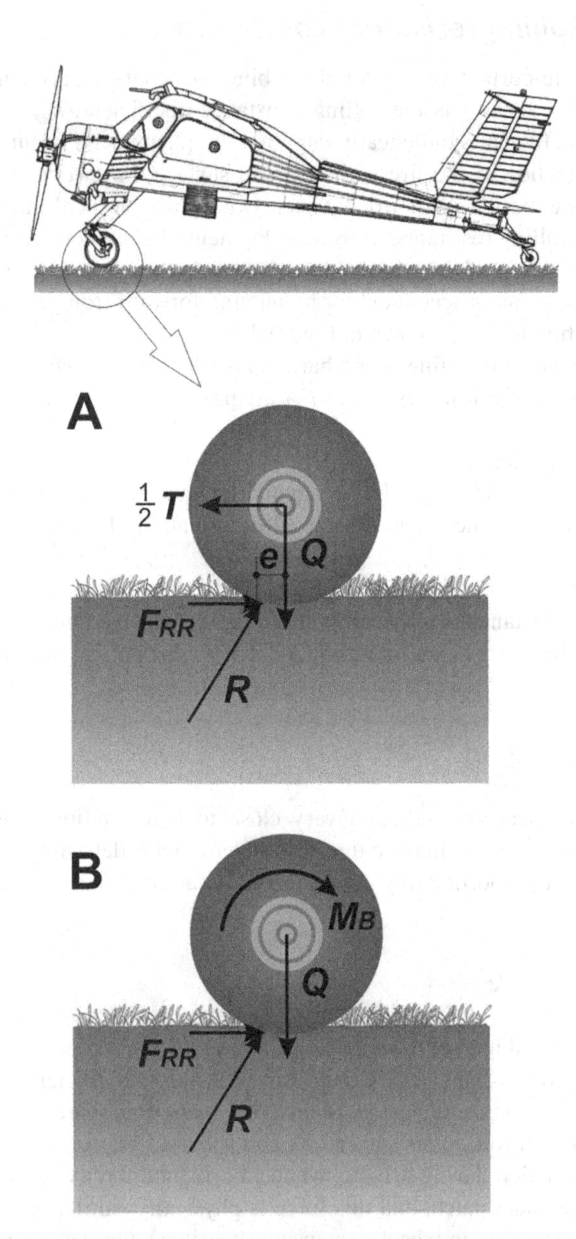

FIGURE 2.9 Forces and moments acting on the landing gear wheel A) during the takeoff run and B) during the landing roll with braking.

The value of the rolling resistance coefficient is dependent upon the speed of motion, and this dependency is described as given here:

$$k_{RR} = k_{R0} + k_{R1}\frac{V}{100} + k_{R4}\left(\frac{V}{100}\right)^4 \tag{2.15}$$

where k_{R0}, k_{R1}, and k_{R4} are coefficients determined experimentally.

The k_{R0} coefficient describes rolling resistance of a tire at speeds near zero, while the k_{R1} and k_{R4} are the coefficients that represent an increase of rolling resistance of a tire at higher speeds of up to 100 km/h and above 100 km/h, respectively. This approach, however, is true only for automotive tires on hard surfaces [9].

The rolling resistance coefficient of a wheel on a hard surface can be determined experimentally by measuring the rolling resistance in a so-called coast-down test or in a tow test.

A typical soft surface (grass, soil, or snow) differs from the aforementioned with an additional resistance component resulting from surface deformation in general [10, 11]. For low speeds, this component can be determined on the basis of Bekker's model [12], but a solution for high speeds requires other methods [13, 14].

Shoop et al. (1999) have concluded that the total rolling resistance on an aircraft on loose soil is due to the internal mechanical resistance (landing gears, wheels suspensions, bearings), aircraft aerodynamic drag, low-speed compaction of the soil, and high-speed resistance due to soil drag on the wheels and spray drag on the body of the aircraft. The soil compaction term was determined with an instrumented vehicle, and the speed-related term was estimated in real size flight tests in which a C-17 transport aircraft was used [15].

After Shoop et al. (1999) rolling resistance F_{RR}^S of a wheel on a soft surface at high speed consists of three components [15]:

$$F_{RR}^S = F_{RR}^H + F_{LS}^S + F_{HS}^S \tag{2.16}$$

where F_{RR}^H is the rolling resistance on hard surface, F_{LS}^S is the resistance due to deformation of soil (or other soft surface) at low speed, and F_{HS}^S is the resistance at higher speeds and consists of the drag of the wheel through the soil, as well as the drag caused by spray of loose soil (or snow) against the aircraft and landing gear.

Rolling resistance attributable to surface deformation at low speed can be determined in tow tests or by the use of specifically instrumented vehicles. Similarly, resistance at high speed can be determined experimentally. There are documented experimental data on high-speed rolling resistance of contaminated airfields [16]; however, data for rolling resistance on unpaved, grassy airfields at various conditions are missing. Yes, the data on the rolling resistance coefficient for the grass surface are cited; however, there is an approximate percentage increase in the coefficient value [17].

2.3.2.4 Methods for rolling resistance measurement

Pull test method

In research praxis, several methods for the measurement of k_{RR} are used. Perhaps one of the simplest is the tow test method. This method is often used for the determination of rolling resistance coefficient in automotive research and testing. A second vehicle tows the test vehicle, and the force needed to pull the test vehicle is measured with the use of a load cell. This method allows the identification of low-speed components of the rolling resistance.

Concluding, the pull test method is suitable for various surfaces with no limitations, but it is not possible for high-speed tests.

Instrumented vehicle method

Another is the instrumented vehicle method. In this method, rolling resistance coefficient is determined with the use of an instrumented ground vehicle. The method, therefore, allows to determine the rolling resistance coefficient for the instrumented vehicle's tires, not aircraft tires. More details on this method are included in Chapter 4.

Portable wheel tester

Another method is to measure the rolling resistance with the use of a portable wheel tester. This type of device is often used by highway construction engineers to test the surface. Different versions of a wheel tester are known, the device basically measuring the rolling resistance force, the traction force (grip), and possibly the distance. It is portable, with dimensions that allow it to be placed in the trunk of a passenger car. The portable wheel tester developed at the Lublin University of Technology has been presented in Chapter 4, while results of measurement are included in Chapter 5.

Flight test method

This method has been developed and used by Pytka et al. (2006) and Pytka (2014). The idea of the method is to measure aircraft ground speeds during takeoff runs and, based on this data, to calculate the acceleration of the entire airplane, then to obtain the rolling resistance coefficient as a result of inverse calculations [2, 18, 20]. Details on this method as well as the results obtained are included in Chapters 4 and 5.

2.3.3 Effect of wind

Headwind shortens the takeoff or landing distance by a factor as follows:

$$\left(1-\frac{V_w}{V_R}\right)^2 \tag{2.17}$$

where V_w is the frontal component of wind speed and V_R is aircraft's rotation speed.

It is worth mentioning that the official FAA prescription accounts for this by lowering the exponent 2 down to 1.85. In addition, it should be remembered that determining the headwind component is difficult unless we have data from the airport weather service. Having the given value of the wind vector, knowing the direction of the runway, the component is determined using the crosswind component graph included in the flight manual.

Crosswind makes it difficult to maintain direction and affects the stability of the movement during the take-up run. The pilot's technique and experience play a decisive role. Note that the maximum crosswind vector value is limited in the airplane manual.

2.3.4 Effect of density altitude

For a piston engine, the horsepower is directly proportional to the massflow of air into the intake manifold. Massflow into the reciprocating engine is affected by the outside air density and intake manifold pressure. It is often said about density altitude, which is a theoretical height, at which the air density is equal to the standard density according to ISA (International Standard Atmosphere). Constant values in the reference atmosphere are assumed for the altitude equal to sea level. They are temperature of 15°C and pressure of 1013,25 hPa. Density altitude is an abstract parameter that allows, among other things, to estimate the performance of an airplane with changing temperature and pressure values. A rule of thumb is as follows: the density altitude increases by 120 ft for each 1°C increase in temperature in the reference atmosphere. An example: If the temperature is 30°C at sea level, the density altitude is 1800 ft.

At this point, it should be remembered how it is designed and how a naturally aspirated (without supercharger) piston engine works. Well, this type of engine has the property that the output power is dependent on the ambient air pressure: the lower the air pressure, the lower the shaft power. This fact is of great importance for ground performance as well as flight performance.

A simple conclusion is that the higher the height of the airport, the longer the takeoff run. It is assumed that for airplanes powered by piston engines (without supercharging), an increase in density altitude by 1,000 ft increases the takeoff run length by approximately 10%.

The decrease of shaft power P output with diminishing air density at altitude is expressed as:

$$P(\varphi)=\Phi(\varphi;\gamma)\times P(\varphi=1) \tag{2.18}$$

where φ is relative atmospheric density and γ is the engine ratio of power lost to friction.

The factor,

$$\Phi(\varphi;\gamma)=\frac{\varphi-\gamma}{1-\gamma}, \tag{2.19}$$

was originally arrived at by Gagg and Farrar [19].

Typical values of γ are close to about 0.12. For example, the value of γ for the Lycoming O-320-D2J engine (a popular engine type for light GA airplanes) is 0.1137.

2.4 CONCLUSION

This chapter presents an introduction to airfield performance of airplanes on grass runway with a particular emphasis on the methods of analysis and measurement of takeoff and landing distances, stability during ground roll, and the influence of significant factors. The main goal was to analyze the literature in terms of searching for optimal analytical and experimental methods. As there are currently no special methods and procedures for testing both the parameters of the state of grass airfields and the performance of the aircraft on a grass runway, an attempt was made to search for such methods on the basis of the state of knowledge and technology of road engineering and automotive technology, which appear to be promising for this research.

The general conclusion that can be drawn from the aforementioned analyses and test results is that the optimal methods of testing the ground performance of the aircraft on a grass airfield should take into account the mechanics of the wheel – deformable substrate system.

REFERENCES

1. Rowe, R.S., Hegedus, E. *Drag Coefficients of Locomotion over Viscous Soils*. Department of the Army, Ordnance Tank – Automotive Command, Land Locomotion Laboratory, Report No. 54, 1959
2. Pytka J., Tarkowski, P., Kupicz, W. A research of vehicle stability on deformable surfaces. *Eksploatacja i Niezawodnosc – Maintenance and Reliability*, 2013, 15(3), 289–294
3. Raymer, D. *Aircraft Design: A Conceptual Approach*. AIAA, Washington, DC, 1989
4. Abzug, M.J. Directional stability and control during landing rollout. *Journal of Aircraft*, May–June 1999, 36(3). Crenshaw, B. Soil –wheel interaction at high speed. *Journal of Terramechanics*, 1972, 8(3), 71–88
5. Pytka, J., Tarkowski, P., Dąbrowski, J., Zając, M., Konstankiewicz, K., Karczewski, L. Determining the ground roll distance of an aircraft on unsurfaced airfield. *Proc. 10th European Conference of the ISTVS*, Budapest, Hungary, October 2006
6. Stinton, D. 1998. *Flying Qualities and Flight Testing of the Aeroplane*. Blackwell Science, Oxford, UK
7. van Es, G.W.H. *Rolling Resistance of Aircraft Tires in Dry Snow*. National Aerospace Laboratory NLR, Report No. NLR-TR-98165, 1998. van Es G.W.H. Method for predicting the rolling resistance of aircraft tires in dry snow. *Journal of Aircraft*, 1999, 36(5), 762–768
8. www.sarsys-asft.com (8 September 2022)
9. Pytka, J., Tarkowski, P., Dąbrowski, J., Bartler, S., Kalinowski, M., Konstankiewicz, K. Soil stress and deformation determination under a landing airplane on an unsurfaced airfield. *Journal of Terramechanics*, 2004, 40, 255–269
10. Lowry, J.T. *Performance of Light Aircraft*. AIAA Publishing, Reston, VA, 1999
11. Shoop, S.A., Richmond, P.W., Eaton, R.A. Estimating rolling friction of loose till for aircraft take-off on dirt runways. *Proc. 13th International Conference of the ISTVS*, Munich, Germany, September 1999. Part I, pp. 421–426.
12. Howland, H.J. Soil inertia in wheel-soil interaction. *Journal of Terramechanics*, 1973, 10(3), 47–65. Raymer, D. *Aircraft Design. A Conceptual Approach*. American Institute of Aeronautics and Astronautics, Inc., Washington, DC, 1992.
13. Eden, P.E. Tractors, trials & lunchtimes. Testing the PC-24 jet's extraordinary capability to operate from unpaved surfaces presented unique challenges. *Aerospace Testing International*, September 2021, pp. 42–48
14. Konstankiewicz, K. The experimental verification of the mechanical model of soil medium. *Proc. of the III European ISTVS Conference*, Warsaw, Poland, 1986, pp. 34–42.
15. Stanton, B.E. Power ranger. Scrappy bush plane built for high payload, endurance and adventure. *Sport Aviation*, January 2022, 71(1), 40–50
16. Gibbesch, A. 2003. Reifen-Boden Interaktion von Flugzeugen auf Nachgeibigen Landebahnen bei hohen Geschwindigkeiten. *Proc. Luft-und Raumfahrt Kongress*, DLR-Deutsches Zentrum fur Luft- und Raumfahrt, pp. 2079–2085
17. Technical Issue Paper EASA-005 – Unusual landing operations. Certification Authorities for Large Transport Aircraft (CATA), November 2018

18. Pytka, J., Budzyński, P., Józwik, J., Michałowska, J., Tofil, A., Łyszczyk, T., Błażejczak, D. Application of GNSS/INS and an optical sensor for determining airplane takeoff and landing performance on a grassy airfield. *Sensors*, 2019, 19, 5492. doi:10.3390/s19245492
19. Mitschke, M., Wallentowitz, H. *Dynamik der Kraftfahrzeuge*. Springer Verlag, Berlin Heidelberg, Germany, 2004
20. Pytka, J.A. Identification of rolling resistance coefficients for aircraft tires on unsurfaced airfields. *Journal of Aircraft, AIAA Journal of Aircraft*, 2014, 51(2), 353–360

Modeling of airfield performance of airplane on a grass runway

3

3.1 INTRODUCTION

Terramechanics is a science grounded in vehicle mobility and soil mechanics, and its primary purpose is to determine wheel (track, ski) performance on soft, deformable surface. A traditional approach to off-road traction modeling lies in the basis of the Coulomb–Mohr hypothesis. This oldest model of soil mechanics assumes that the soil is a material with internal friction. A sudden breakdown of soil structure occurs when a certain level of stress has been reached. The limiting level of stress (shearing stress) is expressed by two basic parameters: soil cohesion and angle of internal friction. Based on this approach, the driving force (traction) on a wheel can be predicted; a number of models have used it as a basic relationship. The simplicity of the Coulomb–Mohr model, however, is accompanied by substantial drawbacks, such as the lack of information on the stress–strain behavior and the fact that it doesn't consider time. Another basic theory for off-road performance prediction is Bekker's model of pressure–sinkage behavior. This model calculates rolling resistance and drawbar pull for vehicles over different soils and uses a sinkage coefficient and two sinkage moduli, which can be determined with a specific test equipment, called the *bevameter* (**Be**kker's **va**lue **meter**) [1, 2]. This theory had been used by Kuchinka, who derived equations for takeoff and landing distances for an aircraft operating on any soft surface, whose Bekker's coefficients are known. Predicted airfield performance had

DOI: 10.1201/9781003312765-3

been compared with flight test results, conducted on sand, sod, and clay soil fields by Lockheed-Georgia Company [3].

Advanced soft surface modeling includes an approach in which soil or snow is treated as a Drucker–Prager's material with plasticity limits. The finite element method (FEM) enables high-fidelity simulations, which gave stress distribution, surface deflection, and wheel motion resistance [4–7].

3.2 WHEEL–SOIL INTERACTION MODELING

Forces acting between a wheel and a surface as well as soil stress and deformation belong to the basic engineering as well as research problems in terramechanics. Off-road performance of a wheel differs from that a classical vehicle mechanics (i.e., on a hardened surface) mainly because of soil or turf deformability and compressibility. Vertical deformations of surface (compression) under a running wheel lead to an increase of motion resistance, caused by a loss of driving energy due to higher rolling drag. Similarly, longitudinal deformations (shearing) cause less driving action as a result of wheel slip and a loss of friction between a wheel and a soil surface. Those effects depend on the soil surface and its actual conditions, so off-road performance may vary within a wide range for a given wheel.

There are known numerous methods for off-road performance prediction. An archetypical solution to the problem of wheel action on a deformable surface was derived from the aforementioned Coulomb's yield criterion, which determines the shear resistance of a plane in a granular body (such as soil) as a function of both internal friction and cohesion. Based on this law, the driving force of a wheel on a soil surface can be predicted. Many models were developed based on this hypothesis, and some of them are still in use.

The aforementioned approach is physically well based, and models derived give good results. The major problem is obtaining soil stress values, which are needed for force calculations. Wulfsohn and Upadhyaya (1992) predicted traction in the soil profile on the basis of 2D and 3D representations of the dynamic wheel–soil contact area. These authors assumed a semi-logarithmic porosity–stress relationship for the determination of pressure distribution along the contact path [8]. Muro (1993) presented an analytical method for predicting the traction of a rigid wheel on soft ground in which the normal stress and shear resistance applied around the peripheral contact part of the wheel were calculated by the use of a dynamic pressure–sinkage curve, obtained in plate loading and unloading tests [9].

The traction prediction equation introduced by Godbole et al. (1993) uses the soil deformation modulus and physical properties of tires as input parameters. They assumed a nonlinear shear stress distribution and change in the values of soil deformation modulus with normal stress. This method, however, requires complex calculations of contact path area [10].

Wanjii et al. (1997) developed an algorithm for the calculation of tractive forces with respect to soil stress distribution derived from Maxwell's model. Soil material parameters needed for calculations are the soil elasticity modulus, which can be determined in penetration tests with the use of a standard loading plate. The only unknown terms required in this algorithm were horizontal distances of bottom dead center of the wheel to initial and rear-end wheel–soil contact points. A good correlation between model and experiment results was obtained for a sandy-loamy soil [11].

Another approach to the problem of wheel performance on deformable surface has been proposed by Pytka (2013). Soil stresses have been measured in various conditions or wheel function modes (rolling, driving, turning, etc.), and tractive forces have been calculated on the basis of measured stresses [12].

3.3 EFFECT OF SOIL DEFORMATION RATE

Basically, soil deformation rate has a significant effect on wheel–soil interaction mechanics and consequently on airplane wheel performance on a grass runway. A research into deformation rate effects on the wheel–soil interaction is a classic terramechanical problem and has been undertaken many times. Rowe and Hegedus (1954) studied rolling resistance in soils with high Moisture Content (MC). They proposed an equation in which the rolling resistance results from the viscosity and static pressure as well as the velocity pressure (dynamic resistance) [36]. Crenshaw (1972) developed a method of determining rolling resistance assuming that the dynamic sinkage of a wheel is a function of the tire diameter, deflection, cross-section height, soil cone index (CI), and some empirical coefficients. These coefficients are determined on the basis of soil data and by measurements on an instrumented aircraft [14]. Hovland (1973) developed a method that takes into account the influence of inertial forces in moving soil. The model developed on these assumptions made it possible to estimate the speed at which the wheel sinkage is so large that it immobilizes a vehicle or a plane. The relationship between the soil lift force generated and the ground speed of the Cessna 150 airplane was proposed [15]. Gibbesch (2003) presented a method for calculating traction forces on aircraft wheels running on soft ground. This method assumes the influence of soil elasticity and viscosity upon contact pressures. The surface of the runway was described using a rheological model. The method was used to determine the vertical forces acting on the wheels of a heavy cargo airplane [16]. Coutermarsh (2007) conducted experimental studies of the influence of deformation velocity on the wheel performance. The rolling resistance of a car wheel with a 235/75R15 tire inflated to 241 kPa, in dry, loose sand, at driving speeds from 2 to 18 m/s, was measured. It was noticed that the rolling resistance increased with speed [17]. This result is similar to those obtained by Pytka (2014), who has measured ground speed of an airplane during takeoff ground roll and has thereof derived rolling resistance coefficients for increasing velocity [18].

One approach to the study of the effect of deformation velocity in terramechanics is the rheology method. Typically, rheological models are feasible to predict soil behavior in civil engineering problems (foundations, etc.), where deformation rates are relatively low. An approach that could give promising results in wheel–soil interactions was proposed by Pukos (1990) who investigated the effects of dynamical loading on soil at various deformation rates, similar to those caused by agricultural and off-road vehicles [19]. He proved that the widely used rheological models were not suitable for describing soil stress–strain processes at high deformation rates, and he proposed a new model that takes into account time effects upon soil elasticity and viscosity. Assuming that in moist soil, deformation rate mostly affects viscosity, a typical rheological model (Maxwell or "a three-element" model) can be expressed in the following differential equation:

$$E_1 E_2 \varepsilon(t) + (E_1 + E_2)\frac{A\sigma(t)}{\sinh(B\sigma(t))}\frac{d\varepsilon}{dt} = E_1\sigma(t) + \frac{A\sigma(t)}{\sinh(B\sigma(t))}\frac{d\sigma}{dt} \tag{3.1}$$

where σ is the soil stress, ε is soil strain, E_1, and E_2 are soil moduli of elasticity, and v is the soil viscosity.

Eq. (3.1) has been solved numerically, since *sinh* is a nonelementary function. The model was experimentally examined, and a new theory in statistical mechanics was created on the basis of Eyring's assumption that soil is partially a Newtonian fluid in which viscosity is temperature dependent. Pytka and Tarkowski (2011) have developed a nonlinear dynamic model of wheel–soil system, on the basis of Eyring–Pukos viscosity for soils [20]. The model gave satisfactory results for a turning wheel on a sandy soil surface.

In the case of airplanes operating on soft, grassy surface, the most significant problem is that of increasing rolling resistance by soft surface effect. The Joint Aviation Authority (JAA) has issued the Advisory Material Joint AMJ 25X1591, which gave information and recommendations as well as methods for calculation of rolling resistance on unsurfaced airfields. Van Es (1999) developed a method for the determination of aircraft wheel rolling resistance in snow [21]. It was assumed that high-speed compaction of a compressible material (soil, snow, grass) causes a resistance related to the speed with which the material is compressed. Moreover, it was suggested that this resistance is related to the increase in kinetic energy of the particles of the contaminant (snow, sand, etc.). As a result of this study, a method was developed that was based on two terms:

- drag due to compression and
- drag due to motion.

One certain limitation of the method is that it is not recommended for wet snow or wet soil. Shoop et al. (1999) have developed a method to estimate the rolling resistance of the airfield to predict takeoff distance [22]. It was concluded the total rolling resistance on an aircraft on loose soil is due to the internal mechanical resistance (landing gears, wheels' suspensions, bearings), aircraft aerodynamic drag, low-speed compaction of the soil, and high-speed resistance due to soil compaction drag acting on wheels and spray drag acting on the body of the aircraft. The soil compaction drag was measured with the use of an instrumented vehicle. The spray drag component was estimated in real-size flight tests with the use of a C-17 military cargo aircraft. Generally, rolling resistance $F_{RR}{}^{S}$ of a wheel on a soft surface at a high speed can be described as a sum of three components by means of the following equation:

$$F_{RR}^{S} = F_{RR}^{H} + F_{LS}^{S} + F_{HS}^{S} \tag{3.2}$$

where $F_{RR}{}^{H}$ is the rolling resistance on hard surface, $F_{LS}{}^{S}$ is the resistance due to deformation of soil (or other soft surface) at low speed, $F_{HS}{}^{S}$ = is the resistance at higher speeds and consists of the drag of the wheel through the soil, as well as the drag caused by spray of loose soil (or snow) against the aircraft and landing gear [22].

3.4 EFFECT OF VEGETATION

Runway vegetation affects the performance of landing gear and, consequently, airplane performance during taxiing, takeoff, and landing. Generally, the vegetation can be divided into the green part (above-ground) and the roots (underground part). The green part of vegetation comes into direct contact with the tire and causes a loss of traction parameters of the wheel. One mechanism is that an additional drag force acts on wheels at rolling and that this effect is due to grass blades' bending (or fracture) as well as due to the compaction of grass under a rolling tire. Another effect is less friction between the tire and the grass, which results in less braking intensity, especially when grass is moist. Another effect of the presence of grass on the runway is the possible loss of motion stability during takeoff or landing ground roll, especially at high speed. One positive effect of vegetation on the runway is lower dynamic loads at touchdown on the landing gear wheels and the entire structure of the aircraft, as well as lower tire tread wear. Primary models respected grass as a homogenous contaminant on runway surface and describe it by means of and increase in rolling resistance coefficient or decrease in braking friction coefficient. In this approach, grass length and grass moisture were taken into account.

Stinton (1998) proposed a simple classification, distinguishing between short dry and short wet grass and tall dry and tall wet grass. For all types, he provided approximate values of the rolling resistance and braking friction coefficients. Stinton's classification is based on the following grass length scale: fresh cut, short, typical summer airfield, tall but useful, lush; too long with appropriate blade lengths in inches (0–2, 2–4, 4–6, 6–10, and 10–12). A caution is that the length of the blades of grass is not equal to the height of standing (the grass may fall under its own weight [23]. However, there was no correlation to wheel performance.

An approach based on the analysis of the grass' physical properties, in particular its mechanical strength, may yield better results. According to agricultural handbooks, grass exhibits measurable physical properties such as bending and compaction strength and friction [24, 25]. For a given grass species, there is a strong relation between mechanical properties of grass blades and its length, diameter, and moisture. These properties are taken into account in agricultural equipment design. Kanafojski (1980) determined the bending strength of a single grass blade with the use of the following equation [24]:

$$EJ = \frac{FL^3}{3f} \tag{3.3}$$

where E is Young modulus, J is mass inertia factor, F is force applied to a grass blade, L is distance between the ground and a point where the bending force is applied, and f is blade deflection.

The term EJ in Ncm^2 varies between 1.5×10^2 and 10^3 for grass leaves and between 10^4 and 10^5 for grass blades. Kanafojski gives also sample values of friction coefficients for grass blades in contact with various materials (steel, rubber, and wood) [24]. The same reference examines the effect of velocity on the grass–steel friction. Within the speed range of 10 to 300 mm/s, friction coefficient changes from 0.15 to 0.30. The relationship of speed–friction demonstrates exponential character [24].

Cenek et al. (2005) investigated the friction between the tread of a car tire and a grassy surface. The aim of the research was to obtain the value of the μ coefficient, mainly to improve driving safety on roads occasionally covered with vegetation (New Zealand). Two measurement methods were used, which were tire dragging and braking of a braked vehicle. The μ values obtained in both tests differed significantly (0.59 versus 0.17) [26].

Białczyk et al. (2006) measured the net traction force of a horticultural tractor equipped with two types of tires: normal and grass tires. Measurements were made for four different species of grass. The results of the experiment show that the values of μ were relatively high for the studied grass areas (between 0.69 and 0.91) [27]. A significant influence of grass species has been observed.

The mechanism of the influence of the roots on the road wheel is slightly different. Underground parts of grass plants and their roots form a complex and strong structure, and, together with green parts, build a lawn (or a sod). It is expected that the wheel–soil interaction will improve, especially by increasing the bearing capacity. Effect of roots on soil shear strength has been researched by Yoshida and Adachi (2001). It was shown that the soil samples with roots (rice plant roots) exhibited much higher shear strength than those without roots [28]. Pirnazarov et al. (2013) and Pirnazarov and Sellgren (2015) developed a new laboratory shear test for tree-root-reinforced soils, applying shear at two levels and considering different root arrangements. This simulates wheel–soil interaction where shear occurs in two vertical planes parallel to the direction of applied wheel load and perpendicular to the root layer (for trees). The main purpose was to support the process for the development of powerful and more environmentally friendly new-generation forest machines [29, 30].

From the point of view of terramechanics and the performance of the road wheel or the entire vehicle, the most detailed and complete analysis of the impact of vegetation was carried out by Shoop et al. (2015) [2] and Wieder and Shoop (2018) [31]. The influence of parameters describing biomass (leaf weight, leaf length, leaf surface area, leaf average diameter, leaf dry weight as well as roots parameters) on motion resistance and net traction was experimentally determined. The highest value of the R^2 parameter was obtained for the relationships of leaf surface area-net traction as well as root length-net traction. Measurements were carried out on two different soil substrates – sandy and clay soils. It was found that the influence of vegetation for both examined soils is quite different. Results showed that biomass had a positive benefit on sandy soils with an increase in biomass-increasing net traction. On the other hand, for clay soil, a decrease in motion resistance with an increased root diameter was observed. Initial studies showed that the increased biomass had a generally positive increase in net traction and a decrease in motion resistance [2].

3.5 EFFECT OF SOIL MOISTURE

Based on both textbooks [1, 12, 32] and current research works [33, 13], it can be concluded that MC effect upon mechanical properties of soils is very significant. Consequently, this factor may also play an important role for airplane performance on grass runways. Two mechanisms can be distinguished. First, the braking friction coefficient, μ, becomes low due to the reduced friction between the tire and the blades of grass. Second, the bearing capacity of soil underneath alters. Consequently, in moist conditions, rolling resistance increases and the wheel sinks deeply. There is a significant influence of soil type on the sensitivity to MC. A typical loess soil in dry conditions is very hard while it tends to deform plastically at MC values between 12% and 20%. Upon reaching a critical MC, the loess becomes plastic and flowable. Sandy soils exhibit different behavior. In dry conditions, sand is a granular material without cohesion. At low MC values, the so-called apparent cohesion appears, which causes a light but noticeable increase in mechanical strength. Further growth of MC in sand leads to a significant change in the mechanical properties. Then the soil becomes semifluid, with a noticeable effect of the granular structure of sand.

The effect of MC on trafficability of a soil can be expressed by the following empirical equation [33]:

$$\ln(RCI) = 4.605 + \frac{2.123 + 0.008C - 0.693\ln(MC)}{0.0149 + 0.002C} \tag{3.4}$$

where RCI is the rating cone index, C is the percentage of clay in the soil tested, and MC is the soil moisture content, %.

MC in a grass airfield substrate changes due to weather conditions. Precipitations irrigate grass field, and sun operation, wind, and soil water fluxes dry it. The dynamics of soil moisture was the subject of research conducted by the present author. Pytka et al. (2019) have measured MC in four different natural surfaces, which were three soils (sand, loess, and moschus) as well as in a turf throughout the entire year 2007, and in exclusive freezing periods. Measurements were done two times each week, with the use of a handheld Time Domain Reflectometry (TDR) meter. The highest dynamics was observed in spring–summer months [34]. Other measurements were done for a grass runway of Radawiec Airfield after intensive rainfall. The MC data was collected in various time intervals for the five following days after the rainfall. The highest MC value was as high as 37%, and the tendency to decrease was rather low. After five days of weather with cloud cover 3/8, winds of 3–5 m/s, the MC decreased to approximately 28%. Another significant decrease in MC value was observed after mowing the grass – soil moisture decreased to 25%.

3.6 SYNTHESIS OF THE MODEL

When constructing the airfield performance model on a grass runway, it was decided that the model should be as simple as possible due to the utilitarian goal of practical implementation in the form of an online application for pilots and users of grass airfields [35]. Therefore, an empirical model based on the CI was selected.

It was assumed that the model will be based on the following algorithm:

1. Determination of tractive forces, in particular the rolling resistance force and the braking friction force, based on the CI value of the runway surface
2. Determination of the k_{RR} and μ coefficients
3. Determination of soil moisture and biomass effects
4. Determination of the influence of external conditions (wind and density altitude)

3.6.1 Influence of the soil substrate of the grass runway

A cone penetrometer will be used to determine the influence of the soil substrate condition, and the tractive forces, that is, the rolling resistance force and the braking force, will be determined on the basis of the penetration resistance readings.

3.6.2 Determination of tractive forces

The following equations will be used to determine the forces F_{RR} and F_{BR} operating in the tire–grass runway system:

- rolling resistance force

$$F_{RR} = \frac{1.2W}{(CI \times b \times d)} + 0.04 \tag{3.5}$$

where W is the wheel load, b is the tire width, and d is the tire diameter;

- braking friction force

$$F_{BR} = 0.75\left[\left(1-\exp\left((CI\times b\times d)\frac{0.3i}{\mathrm{W}}\right)\right)\right] \tag{3.6}$$

The application of the aforementioned formulas is correct as long as the soil of the runway is compacted, which is true in most cases. The k_{RR} and μ coefficients are determined, taking into account the vertical load on the airplane wheel.

3.6.3 Modeling the influence of significant factors

3.6.3.1 Soil moisture content

The effect of soil moisture is accounted for by the following equation describing the corrected coefficient of braking friction μ_{grass}:

$$\mu_{grass} = \mu\left(1-\exp\left(MC\right)\right) \tag{3.7}$$

where MC is the soil moisture content.

3.6.3.2 Vegetation

The following relationship was proposed, which takes into account the effect of green parts of vegetation (green parts of grass):

$$k_{RR}^{grass} = k_{RR}e^{m_G} \tag{3.8}$$

where k_{RR}^{grass} is the rolling resistance of a tire on grass runway and m_G is the biomass, measured in kg/m^2.

3.6.3.3 Wind and density altitude

The influence of the wind was taken into account as follows:

$$V_f^{wind} = V_f - V_w \tag{3.9}$$

where V_f^{wind} is the corrected liftoff speed, V_f is the liftoff speed, and V_w is wind speed component parallel to the takeoff direction.

The effect of density altitude is described by the Gagg–Ferrar model, which takes into account the effect of reducing the engine horsepower P of a naturally aspirated piston engine with increasing flight altitude. It is expressed in the following equation:

$$P = P_{SL}\left(\sigma - \frac{1-\sigma}{7.55}\right) \tag{3.10}$$

where P_{SL} is the engine power at the sea level and σ is the air density index, which is:

$$\sigma = \frac{\rho}{\rho_0} \tag{3.11}$$

where ρ is the altitude air density and ρ_0 is the sea-level air density for a given air temperature.

It should be remembered that density altitude affects not only horsepower, but also the aerodynamic performance of the aircraft, for example, liftoff speed, which also indirectly affects ground performance. Hence, density altitude upon liftoff speed V_f is as follows:

$$V_f^{DA} = \frac{1}{\sqrt{\sigma}} V_f \tag{3.12}$$

where V_f^{DA} is the corrected liftoff speed.

Another phenomenon not included here is the so-called ground effect. When a wing nears the ground (takeoff or landing) to about a half of the span, the drag due to lift is substantially reduced. Theoretically, this phenomenon can be explained as a reduction in the induced downwash angle. It can be visualized as a trapping of a "cushion of air" under the wing. This effect can be of interest, especially for a grass runway, the roughness of which may increase the described effect.

3.7 CONCLUDING REMARKS

This chapter provides a brief description of the terramechanical method in relation to the landing gear wheel–grass runway surface. The basic methods of terramechanics were presented, as well as their application in the research of the landing gear performance on the basis of the literature. Own model was proposed, which takes into account important factors such as mechanical strength of soil, soil moisture, vegetation, speed of interaction between the wheel and the runway surface, density altitude, and wind. Parameters describing the aforementioned effects have been introduced. Parameterization of models will be carried out using the identification method, on the basis of the results of flight tests.

REFERENCES

1. Bekker M.G. *Theory of Land Locomotion. The Mechanics of Vehicle Mobility*. University of Michigan Press, Ann Arbor, MI, 1961
2. Shoop S.A., Coutermarsh B., Cary T., Howard H. Quantifying Vegetation Biomass `Impacts on Vehicle Mobility. *Journal of Terramechanics*, Vol. 61, 2015, pp. 63–76
3. Kuchinka A.J. Prediction of Off-Runway Takeoff and Landing Performance. *Journal of Aircraft*, Vol. 3, No. 3, May–June 1966, pp. 213–221
4. Aubel T. FEM Simulation of the Interaction between Elastic Tyre and Soft Soil. *Proceedings 11th ISTVS Conference*, Lake Tahoe, NV, USA, 2005
5. Fervers C.W. Improved FEM Simulation Model for Tire-soil Interaction. *Journal of Terramechanics*, Vol. 41, 2004, pp. 87–100
6. Lee J.H. Finite Element Modeling of Interfacial Forces and Contact Stresses of Pneumatic Tire on Fresh Snow for Combined Longitudinal and Lateral Slips. *Journal of Terramechanics*, Vol. 48, 2011, pp. 171–197
7. Shoop S.A. Finite Element Modeling of Tire-Terrain Interaction. ERDC/CRREL TR-01–16, Cold Regions Research Engineering Laboratory, Hanover, NH, USA, 2001
8. Wulfsohn D., Upadhyaya S. Prediction of Traction and Soil Compaction Using Three-dimensional Soil – tyre Contact Profile. *Journal of Terramechanics*, Vol. 29, No. 6, 1992, pp. 541–553
9. Muro T. Tractive Performance of a Driven Rigid Wheel on Soft Ground Based on the Analysis of Soil – wheel Interaction. *Journal of Terramechanics*, Vol. 30, No. 5, 1993, pp. 351–369
10. Godbole R., Alcock R., Hettiaratchi D. The Prediction of Tractive Performance on Soil Surface. *Journal of Terramechanics*, Vol. 30, No. 6, 1993, pp. 443–459
11. Wanjii S., Hiroma T., Ota Y., Kataoka T. Prediction of Wheel Performance by Analysis of Normal and Tangential Stress Distributions under the Wheel – soil Interface. *Journal of Terramechanics*, Vol. 34, No. 3, 1997, pp. 165–186
12. Pytka J. *Dynamics of Wheel-Soil Systems A Soil Stress and Deformation Based Approach*. Taylor & Francis, Boca Raton, FL, 2013
13. Pytka J., Laskowski J., Tarkowski P. A New Method for Testing and Evaluating Grassy Airfields and its Effects upon Flying Safety. *Scientific Journal of Silesian University of Technology – Series Transport*, Vol. 95, 2017, pp. 171–183

14. Crenshaw B.M. Soil-wheel Interaction for High Speed. *Journal of Terramechanics*, Vol. 8, No. 3, 1972.
15. Hovland H.J. Soil Inertia in Wheel-soil Interaction. *Journal of Terramechanics*, Vol. 10, No. 3, 1973, pp. 47–65.
16. Gibbesch A. Reifen-Boden Interaktion von Flugzeugen auf Nachgeibigen Landebahnen bei hohen Geschwindigkeiten. *Proceedings Luft-und Raumfahrt Kongress*, DLR-Deutsches Zentrum fur Luft- und Raumfahrt, 2003, pp. 2079–2085.
17. Coutermarsh B. Velocity Effect of Vehicle Rolling Resistance in Sand. *Journal of Terramechanics*, Vol. 44, 2007, pp. 275–291.
18. Pytka J. Identification of Rolling Resistance Coefficients for Aircraft Tires on Unsurfaced Airfields. *Journal of Aircraft*, Vol. 51, No. 2, 2014, pp. 353–360
19. Pukos A. *Soil Deformation as Affected by Pores and Grains Distribution*. Problems in Agrophysics Series, No. 61, Institute of Agrophysics PAS, Lublin, Poland, 1990
20. Pytka J., Tarkowski P. A Non-Linear Model for a Turning Wheel on Deformable Surfaces. *17th International Conference of the International Society for Terrain-Vehicle Systems*, Blacksburg, VA, USA, September 18–22, 2011
21. van Es G.H.W. Method for Predicting the Rolling Resistance of Aircraft Tires in Dry Snow. *Journal of Aircraft, AIAA*, Vol. 36, No. 5, September–October 1999, pp. 762–768
22. Shoop S.A., Richmond P.W., Eaton R.A. Estimating Rolling Friction of Loose Till for Aircraft Take-off on Dirt Runways. *Proceedings 13th International Conference of the ISTVS*, Munich, Germany, September 1999, Part I, pp. 421–426.
23. Stinton D. *Flying Qualities and Flight Testing of the Aeroplane*. Blackwell Science, Oxford, UK, 1998
24. Kanafojski C. Theory and Construction of Agricultural Equipment. *PZWRiL, Warsaw*, 1980, pp. 23–36
25. Mohsenin N.N. *Physical Properties of Plant and Animal Materials*. Gordon and Breach Publishers, Amsterdam, The Netherlands, 1986.
26. Cenek P.D. Jamieson N.J. McLarin M.W. Frictional Characteristics of Roadside Grass Types. *International Surface Friction Conference*, Christchurch, New Zealand, 2005
27. Białczyk W., Cudzik A., Czarnecki J., Pieczarka K. Analysis of Traction Properties of Grass Area. *Zeszyty Naukowe, Wroclaw Agricultural University*, No. 545, Wroclaw, Poland, 2006
28. Yoshida S., Adachi K. Effect of Roots on Formation of Internal Fissures in Clayey Paddy Soil during Desiccation. *Journal of the Japanese Society of Soil Physics*, No. 88, 2001, pp. 53–60
29. Pirnazarov A., Wijekoon M., Sellgren U., Lofgren B., Andersson K. Modeling of the Bearing Capacity of Nordic Forest Soil. *Proceedings of the 12th European Regional Conference of the International Society for Terrain-Vehicle Systems (ISTVS)*, Pretoria, South Africa, September 24–27, 2012.
30. Pirnazarov A., Sellgren U., Lofgren B. Development of a Methodology for Predicting the Bearing Capacity of Rooted Soft Soil. *Proceedings of the 7th American Regional Conference of the International Society for Terrain-Vehicle Systems (ISTVS)*, Tampa, FL, November 4–7, 2013.
31. Wieder W.L., Shoop S.A. State of the Knowledge of Vegetation Impact on Soil Strength and Trafficability. *Journal of Terramechanics*, Vol. 78, 2018, pp. 1–14

32. Wong J.Y. *Terramechanics and Off-Road Vehicle Engineering: Terrain Behaviour, Off-Road Vehicle Performance and Design*. Butterworth Heinemann, Oxford, UK, 2009
33. Anderson M.G. On the Applicability of Soil Water Finite Difference Models to Operational Trafficability Models. *Journal of Terramechanics*, Vol. 20, No. ¾, 1983, pp. 139–152
34. Pytka J., Budzyński P., Kamiński M., Łyszczyk T., Józwik J. Application of the TDR Moisture Sensor for Terramechanical Research. *Sensors*, Vol. 19, No. 9, 2019, p. 2116
35. Pytka J., Tarkowski P., Budzyński P., Józwik J. Method for Testing and Evaluation of Grassy Runway Surface. *Journal of Aircraft*, Vol. 54, No. 1, 2017, pp. 229–234
36. Rowe R.S., Hegedus E. *Drag Coefficients of Locomotion over Viscous Soils*. Department of the Army, Ordnance Tank – Automotive Command, Land Locomotion Laboratory, Report No. 54, 1954

Methods 4

4.1 GRASS RUNWAY SURFACE CHARACTERIZATION METHODS

4.1.1 Cone penetrometer – a classic terramechanical instrument

One of the simplest measuring devices, the cone penetrometer, is used to determine the soil strength. Its working part, a conical cone with an angle of 30°, is introduced into the soil by means of a servo drive or manually at a constant speed. The penetration force acting during this use is recorded and is used to determine the so-called Cone Index (CI). Figure 4.1 shows a cone penetrometer, which is very common, and the measurement method using this device is considered as a useful method. The CI that can be determined with a penetrometer in the field is a universal terramechanical number. It shows the current and average mobility conditions. The penetrometer reacts very sensitively to differences in soil moisture and inhomogeneities in the soil. Statistically usable results can only be achieved with a number of measurements – typically 5 to 10 but this can be carried out in practice thanks to the simple and fast measuring method of the penetrometer.

There is a special airfield cone penetrometer, consisting of a 30° cone with a 0.2-square inch base area, which has a range of 0–15 (CBR value of 0 to approximately 18), Similar in design to the trafficability penetrometer. Force is applied to the penetrometer at a rate of ½ to 1 in./sec, with readings taken at 2-in. Increments, up to 24 in., or until a maximum reading of 15 is obtained. Readings from the airfield cone penetrometer are reported as the Airfield Index (AI). Readings from the 0.2-square inch cone trafficability penetrometer must be divided by 20 to obtain the AI; the reading obtained with the 0.5-square inch cone must be divided by 50 to obtain the AI [1, 2].

DOI: 10.1201/9781003312765-4

FIGURE 4.1 Measuring soil strength with the use of a cone penetrometer.

From the measurements taken in the field, the following wheel parameters can be estimated:

- the rolling resistance of a wheel/tire, k_{RR};
- tire–road traction coefficient μ.

The rolling resistance coefficient is calculated from the axle load W and the dimensions of the tire on the basis of McAllister's empirical equation [3]:

$$k_{RR} = \frac{1,2W}{(CI \times b \times d)} + 0,04 \tag{4.1}$$

where: b = tire width, d = tire diameter.

The driving force coefficient can be calculated using the equation determined by Wismer and Luth [4]:

$$\mu = 0,75\left[\left(1 - exp(CI \times b \times d)\frac{i}{F_z}\right)\right] - k_{RR} \tag{4.2}$$

Eq. (4.2) is valid only for a compacted soil and with normal tire pressure.

Shoop conducted a research aimed at determining the relationship between the CI and the CBR index (California Bearing Ratio). The practical significance of these tests results from the use of CBR to assess the strength of ground road surfaces, including runways of grass airfields. On the other hand, the CBR measurement method is more cumbersome, requires more work, and requires the use of a vehicle – the carrier for the CBR measurement device. The results of the penetrometer measurement are easier to obtain, but their use for the assessment of the road or airport pavement requires correlation with the CBR values.

The report by Shoop et al. (2008) presents the research program, the results of field measurements as well as the CI–CBR correlation algorithm. The obtained results made it possible to propose a relationship between CBR and CI, taking into account significant factors [5].

4.1.2 TDR – a handheld instrument for soil moisture measurements

The TDR (Time Domain Reflectometry) is an indirect method for measuring soil water content – the travel time of a high-frequency electromagnetic pulse through the soil. Based on the measured travel time, soil dielectric permittivity constant is calculated, and this constant is related to soil water content. A typical TDR meter consists of a probe with two electrodes that are inserted into soil as an electronic module, typically with a display (see Figure 4.2). The measurement is easy and takes only seconds. When measuring grass runway, it is important to remove grass from the test spot, because green vegetation may affect the final result.

One new solution of the TDR meter for soil consists of a pocket size electronic unit, without the display but wirelessly linked with any personal device (smartphone, etc.), which significantly facilitates data acquisition.

The measurement procedure consists in placing the electrodes of the measuring tip in the soil, to a depth of approx. 10–15 cm, and then wait until the measuring system of the instrument displays the moisture value on the screen, which is given as a percentage. Typically, 10–15 measurements are made at different points in the test field, and the results should be averaged.

4.1.3 Method for characterizing the biomass on the grass runway

An important aspect of airplane performance analysis on a grass airfield is the development of a methodology for taking into account the impact of biomass, in particular the above-ground (green) grass parts. Shoop et al. (2015) and Wieder and Shoop (2018) discuss methods of measuring and analyzing the influence of biomass on vehicle traction as well as on the performance of an off-road vehicle [6, 7].

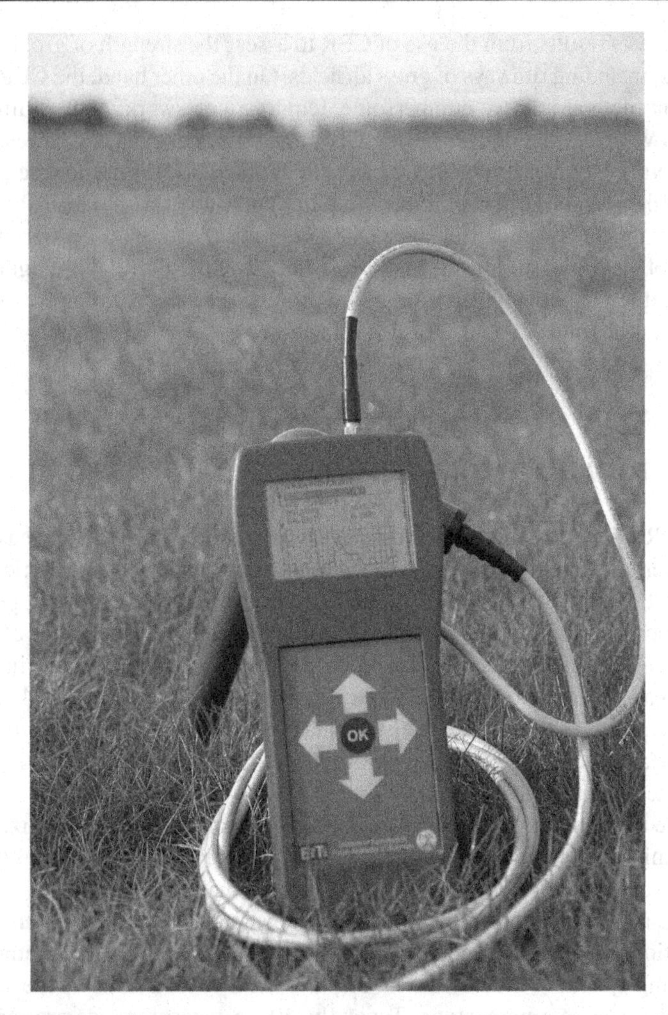

FIGURE 4.2 A TDR meter for soil moisture measurement.

On the basis of these works, the author proposed a simplified method that can be used at grass airfields. The method consists in taking a sample of the ground parts (green) of the grass, measuring their length, weighing them, then drying and weighing again to determine the Moisture Content (MC). Three parameters are obtained:

- the average length of grass blades,
- biomass (mass of the grass blades) related to 1 m^2 of the area, and
- grass moisture.

The method of collecting a grass sample is very simple; it consists in cutting the grass just above the soil surface using any tool (scissors, mower), but it should be done on a specific area, for example, 1 m^2. Then, the cut green grass parts should be weighed, preferably with an accuracy of 1 g, and the biomass is then dried. Finally, the dry grass is weighed and its moisture determined on this basis.

The length of the grass can be measured by any means, both before and after cutting. The photograph (Figure 4.3) shows the taking of biomass samples from the runway of the grass airport (Depułtycze Królewskie, August 2022).

4.2 AIRPLANE PERFORMANCE MEASUREMENT METHODS – GROUND TESTING

4.2.1 Wheel dynamometer for measurements of forces and moments acting on landing gear wheel[1]

Experimental parameterization and the verification of the wheel–soil models on a grass runway can be carried out with the use of specialized measuring methods and devices. The quantities that optimally describe the wheel–ground interaction are the forces and moments in the area of contact between the tire tread and the road surface. In this section, a dynamometer system to measure two force components acting along the longitudinal and vertical axes of the wheel as well as three moments acting around the longitudinal, transversal, and vertical axes has been presented. The dynamometer consists of a sensor unit that is embedded in the wheel hub, a modified rim with a tire as well as a data acquisition and transfer system that enables the measured signals to get transferred wirelessly to a portable computer or another device (smartphone, tablet). The sensor unit was developed on the basis of strain gage measurement technology. The prototype system was calibrated in a stationary test stand then installed on a PZL 104 Wilga 35A for airfield tests. Primary test measurements were performed with the aircraft taxiing at "walking man" speed.

4.2.1.1 Introduction

A wheel dynamometer is a measuring device, typically built – in a road, the wheel of a vehicle enables to measure forces and moments acting on wheel's

FIGURE 4.3 Taking a biomass sample from the runway of the Depułtycze Królewskie airport.

axis, and such measurements can be conducted in real time, on a moving vehicle. In principle, all known wheel dynamometers come from the automotive industry or research, and, typically, a wheel dynamometer is designed as a six-element sensor, on the basis of strain gage or piezoelectric sensor technology.

There are commercially available solutions as well as custom-designed and developed dynamometers that fulfill special requirements. In the automotive research community, a wheel forces/moments reference system based on SAE standards is often used [8–10].

Three perpendicular forces, F_Z, F_X, and F_Y, are acting as follows:

- the F_Z is the vertical force, a sum of the gravity force and dynamical effects during a ride;
- the F_X is a driving or braking force;
- the F_Y is the lateral force in a steered axle, and this force comes from the side slip between the tire and road.

Three moments are acting around the respective axes:

- M_Z is acting around the vertical axis, and in a steered axle this moment is called aligning torque;
- the M_X rotates the wheel around the longitudinal axis and is active in a typical car wheel suspension; and
- the M_Y is a driving or braking moment.

4.2.1.2 Design and development of the wheel dynamometer system

The wheel dynamometer system presented here is capable of measuring five components: wheel forces and moments that act on a landing gear wheel of an airplane in motion on the ground. The complete wheel dynamometer system consists of the following elements:

1) a wheel forces/moments sensor for the measurements of the four components: F_X, F_Z, M_X, and M_Z;
2) a braking moment M_Y sensor;
3) a signal conditioning unit, including five signal amplifiers; and
4) a communication unit for wireless data transfer.

4.2.1.2.1 Wheel forces and moments sensor

The wheel forces and moments sensor utilized strain gage measurement technology and was built as a stator sensor – called a hub dynamometer in automotive technology. The basic function of a strain gage, sensing the strain of a spring element caused by applied load, can be described with the following equation:

$$\frac{\Delta R}{R} = \frac{\Delta \rho}{\rho \epsilon} + 1 + 2v \qquad (4.3)$$

where R is the electrical resistance of the gage, ρ is the specific resistance of the gage material, v is the Poisson's factor for the material of a spring element, and ε is the strain of the spring element due to load.

The term

$$\frac{\Delta\rho}{\rho\epsilon}+1+2v \tag{4.4}$$

is called the gage factor and is denoted as GF.

The spring element of the $F_X - F_Z - M_X - M_Z$ sensor was designed as a monolith, cross-shaped element with four bending beams that connect the inner part – the wheel axis with the outer ring. The strain gages had been placed on the beams and on the outer ring, and they had been connected into the bridge circuit so the resulting equation in a general form that links the measured strains with outgoing signal is as follows:

$$\Delta U=\frac{U}{4}\left(\frac{\Delta R_1}{R}-\frac{\Delta R_2}{R}-\frac{\Delta R_3}{R}+\frac{\Delta R_4}{R}\right) \tag{4.5}$$

where U is the output voltage and R_1, R_2, R_3, and R_4 are the resistances of the strain gages in the bridge circuit.

If the orientation of the beams in the spring element is pure horizontal/vertical, the sensor is capable of measuring the horizontal and vertical wheel forces, F_X and F_Z, as well as the moments, M_X and M_Z. Therefore, it is important that the complete sensor is properly installed in a test airplane.

A critical point is bearing the wheel, since the relatively large size of the stator sensor usually leads to unconventional solutions and major changes in wheel rim construction. Comparing with the rotating dynamometer design, there is no need of transferring data signals from the rotating sensor. The base for the sensing unit was a wheel of a chosen airplane, and it was the PZL 104 Wilga 35A airplane. Since the space available inside the wheel for the sensing element core and the transducers was very tight (approximately 150 mm in diameter × 200 mm width), it was required to design and fabricate a completely new wheel rim and bearing system to fit into the landing gear leg of the test airplane.

The sensor core was made of steel, and strain gages were used as sensing elements. The instrumentation amplifiers, which prepare the outgoing signals for data acquisition, were placed outside the sensor in a small box together with a board of the microprocessor data analysis and transfer system. Figure 4.4 shows the elements of the sensor.

To fix the external diameter of the wheel rim to fit/match with the tire, a custom wheel rim was designed and fabricated of steel. Together with a tire, the wheel was supported with two ball bearings and fastened with a nut. The fabricated wheel rim enables setting the tire inflation pressure, which is very practical during airfield measurements, when the inflation pressure has to be a parameter. The rim should be virtually tested for mechanical strength since this element has to bear loads generated on a wheel during the ground phase of a flight: touchdown, free rolling, and braking. Moreover, because of small place available, the width of the rim's wall could be a maximum of 4 mm. We performed simulations of the deflection of the wheel rim for several different load patterns, taking into account the maximum gross mass of the airplane and different wheel function modes: touchdown, rolling at high speed, braking, etc. Deflections of

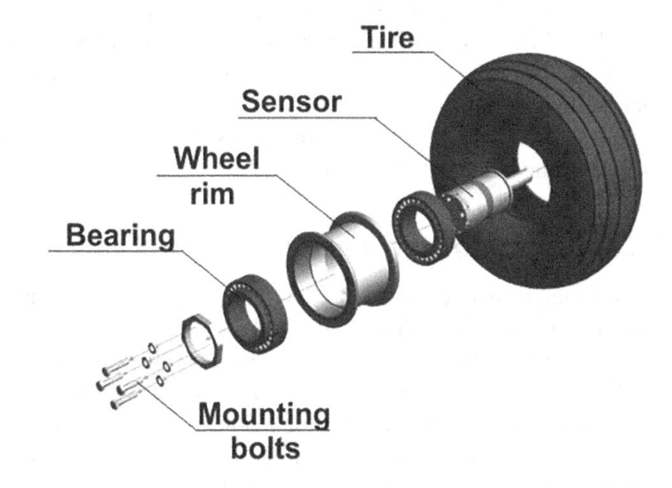

FIGURE 4.4 Elements of the wheel dynamometer.

the wheel rim under loading must not exceed critical values for the bearings, so that they can function properly. Based on those simulations, we have chosen the right material in the sense of mechanical strength of which the wheel rim was fabricated.

4.2.1.2.2 Braking moment sensor

Due to the restricted space inside the wheel rim, we developed a braking moment sensor embedded outside the wheel in the wheel brake. The M_Y moment is measured with the strain gages sealed on the brake jig, which bends proportionally to braking moment.

Two constrictions of defined thickness have been made on two transverse arms of the yoke, which serve as elastic elements. In these places, strain gauges were glued. When the brake is activated, the friction force generates the bending moment acting through the transverse yoke arms. The deflection of the arms in the constrictions is sensed by strain gauges. The solution is shown in Figure 4.5 (a schematic and a photograph).

4.2.1.3 Calibration of the sensor

The X–Y components channels were calibrated on a test stand for obtaining the V-Kn and V-kgm calibration factors. The calibration test stand was adapted from a conventional fatigue tester. Schematic of the calibration is given in Figure 4.6. Note the square fixture attached to the sensor in order to apply loads in two perpendicular directions. For the moment calibration, a force arm has been attached to the sensor. Calibration tests were carried out for every single force channel, and the sensor was positioned in the test rig accordingly for the particular test. The tests were done for ±5.5 kN (F_Z) and for ±1.3 kN (Fx) loads. The moment channels were calibrated within a range of ±450 and 220 kgm. The results of the calibrations are included in Figure 4.7. The results show a very good linearity of the sensor. The force–voltage equations derived from the calibration measurements are as follows:

$$F_Z = 0.4512U_Z + 2.508 \tag{4.6}$$

$$F_X = 1.913U_X + 2.525 \tag{4.7}$$

$$M_Z = 5.461U_{MZ} + 2.507 \tag{4.8}$$

$$M_X = 11.287U_{MX} + 2.507 \tag{4.9}$$

$$M_Y = 0.124U_{MY} + 0.198 \tag{4.10}$$

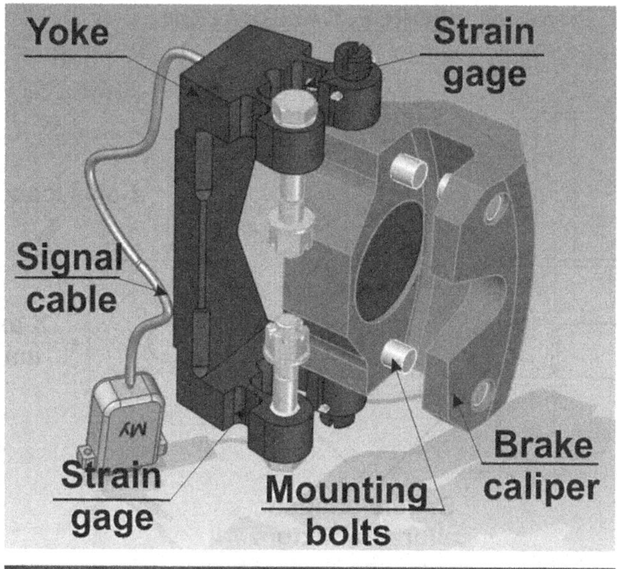

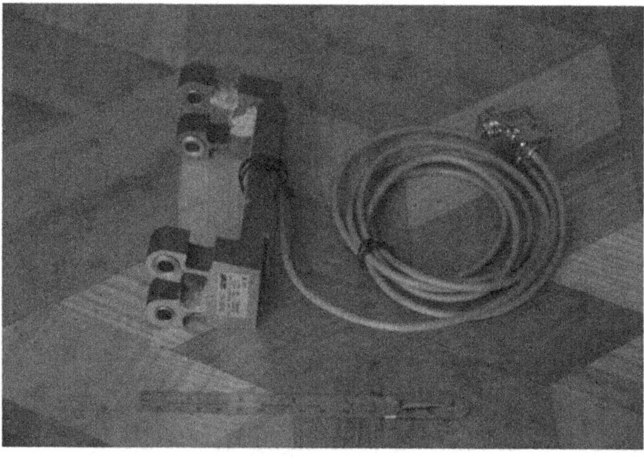

FIGURE 4.5 A schematic of the M_Y moment sensor (upper illustration) and the brake jig with sealed strain gages (lower photograph).

Both sensors were protected against temperature effects by adding the so-called "dead" strain gages in the bridges.

In multi-element force sensors with monolite spring elements, measurements interact with each other. Based on the calibration measurements, these interactions have been determined for the F_X–F_Z–M_X–M_Z sensor. Fractions of the measured values for each combinations of channels inform how much a given channel influences another channel during measurement (under applied

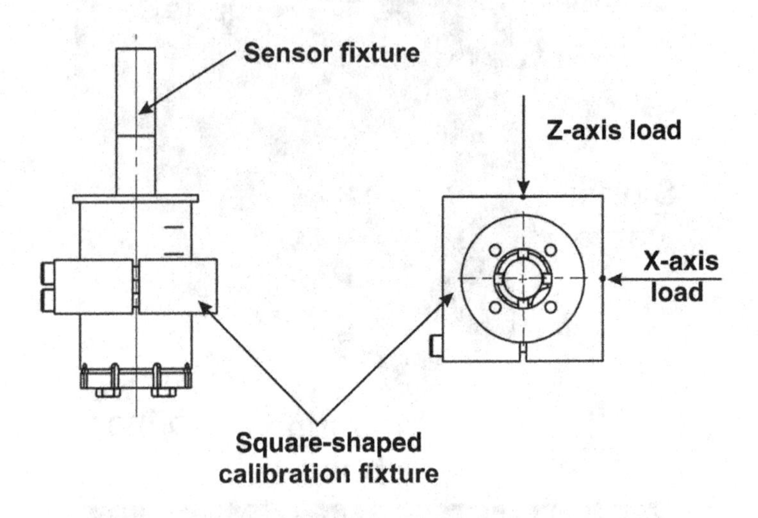

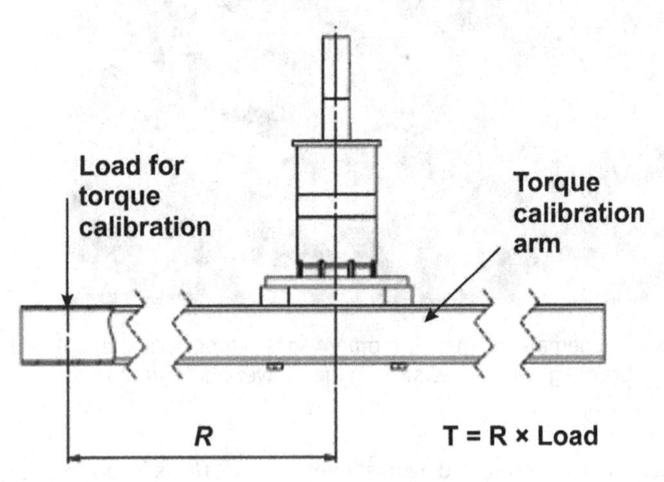

FIGURE 4.6 A schematic of the calibration fixtures.

Source: Reprinted from "Wheel dynamometer system for aircraft landing gear testing"; authors: Pytka J., Józwik J., Budzyński P., Łyszczyk T., Tofil A., Gnapowski E. and Laskowski J.; published in *Measurement* (148) 2019, 106918 (Copyright 2019 with permission from Elsevier)

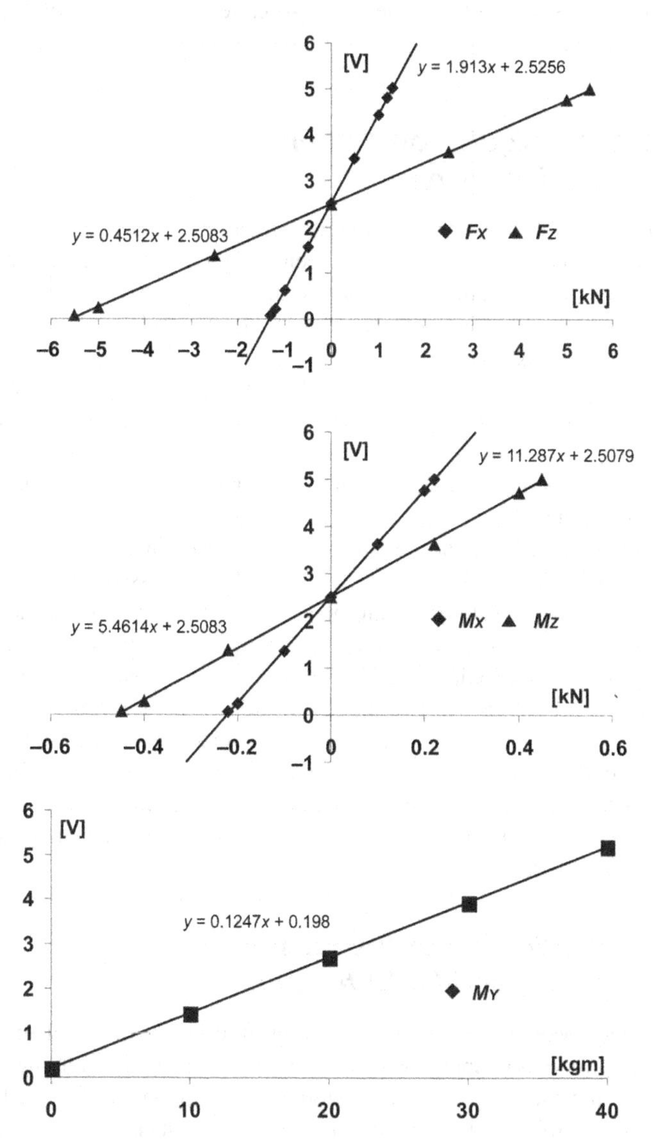

FIGURE 4.7 Results of the calibration measurements.

Source: Reprinted from "Wheel dynamometer system for aircraft landing gear testing", authors: Pytka J., Józwik J., Budzyński P., Łyszczyk T., Tofil A., Gnapowski E. and Laskowski J.; published in *Measurement* (148) 2019, 106918 (copyright 2019 with permission from Elsevier)

load). For the presented dynamometer, the so-called "cross-talks" of the sensor are small, not exceeding 4%.

4.2.1.4 Data acquisition and online transfer system

A miniature data system for transferring signals from the sensor unit to any computing and data storage device has been designed and developed specially for this study. This system includes a microprocessor-based AC for DC data conversion unit and a Bluetooth-based system for wireless transferring the amplified, voltage signals.

The instrumentation amplifiers transform the measured signals into voltage signals within the range of 0 to 4.096 V. The zero point is in 2.048 V so that the system can read both positive and negative forces and moments. The amplifiers have been mounted on separate circuit boards, one each for the five measuring channels.

The data acquisition and transfer system is based on a 16-bit, 8-channel analog-to-digital converter, which performs the successive approximation with the highest data sampling rate of 115 ksps (*kilo samples per second*). Data acquisition is based on a 32-bit controller. Measured signals are transmitted by a Bluetooth Class 2.0 device that ensures long communication range (more than 100 m). The functions of the microcontroller is managing data acquisition, preliminary data analysis, and filtering the digital data. The data acquisition and transfer system has been encapsulated in a hermetic, shock-resistant box, together with the five channel amplifiers. A battery pack of 8.4 V is used as a power supply for the entire measuring system (including the strain gages).

4.2.1.5 Certification of the sensor for ground and flight testing

Since the most valuable data come from real flight testing, one of the future aims of the authors is to apply the presented sensor system into a flying aircraft and performing the measurements of forces and moments during full speed ground tests as well as during takeoff ground roll, touchdown, and landing rollout. To complete this, it is necessary to obtain an official permission of using the wheel dynamometer for test flight. The sensor system in the presented version can be certified for the use on the PZL 104 Wilga 35A aircraft. However, it is possible to reconstruct the sensor system in order to fit to any other light aircraft, and, consequently, the certification could be achieved virtually for other aircrafts. Certification of the wheel dynamometer requires fatigue tests that are planned with cooperation with the Institute of Aviation, Warsaw, where a test rig for the PZL 104 Wilga 35A undercarriage is installed. The certification tests will include drop tests with various initial energies as well as rollout tests on a drum test stand. The certification process has to be conducted under the supervision of the owner of the type certificate for the Wilga aircraft [25].

The complete wheel dynamometer installed on a test airplane, with the electronic unit, have been shown in Figure 4.8.

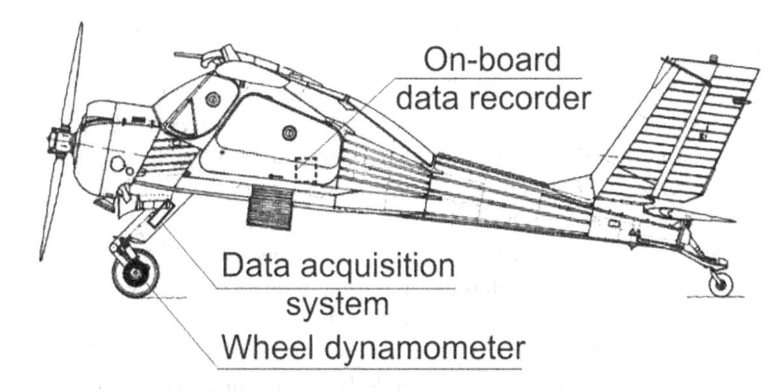

FIGURE 4.8 The wheel dynamometer installed on a test airplane.

4.2.2 Methods for determining the rolling resistance and braking friction

4.2.2.1 Tire–runway tester for the measurement of rolling resistance and braking friction

4.2.2.1.1 The development of the wheel tester

The first portable tester for surface friction and rolling resistance measurement from the Lublin University of Technology has been designed and built in 2016. The tester was a three-wheeled portable device designed for easy and quick operation over a variety of surfaces, roads (asphalt, bitumen, Belgian pavement, etc.), and off-roads (grass, soft soil of smooth surface). In the design, two strain-gage-type transducers have been used to measure horizontal force acting on the axis of a test wheel. The front test wheel together with two rear wheels are of 150 mm diameter, and they have pneumatic tires with air valves in order to maintain inflation pressure. The main frame of the tester supports rear wheels, which rolls freely and a suspension of the test wheel. The suspension integrates the test wheel and its axis, brake disc bearings, and their housings, which are mounted to the strain gage transducers. The suspension functions so that any action (a force or a torque moment) on the test wheel in horizontal plane is reduced to pure longitudinal forces acting 50%/50% on the two transducers. The bearings that support the test wheel axis are rocker type to compensate any nonlinearities that could be the result of assembly as well as the measuring action.

In order to reconstruct wheel–surface interactions, an effect of tire size (width, diameter) has been considered. Regarding the contact pressure as a parameter describing the wheel–surface interactions and disregarding the tread effects, contact area and contact pressure for the test wheel under various loads have been determined at a normal inflation pressure. Wheel load–contact pressure pairs have been taken to construct a relationship which is a nomogram to calculate an additional weight that has to be added to load the test wheel for a given tire. The net weight on the test wheel (without additional ballast) is 3 kg. Sport weight-lifting round weights have been used, and this solution enables to apply the additional load in steps: 5, 10, 15, 20, 25, 30, and 35 kg. Ballast heavier than 50 kg can damage the mechanical structure of the tester and may have significant effects on load cells readings. Based on this simple model, it is possible to determine additional weight for a given tire for which we want to determine the μ or k_{RR} value. As an example, for a tire that generates contact pressure 100 kPa, an additional weight of 30 kg is needed to reconstruct tire–surface interactions and determine the values of μ and k_{RR}.

The software developed in that study used an algorithm that allows to choose a range of *My* data for averaging and determining the f_{RR}. An operator selected the starting and ending point of a range of measured data. The algorithm included a smart tool for searching for maximum values of *My* data, which was used to determine braking friction coefficient μ_B. The results of measurement, f_{RR} and μ_B, were displayed on a screen of the portable device to which the tester was connected (a smartphone, tablet, etc.) [11].

The measuring wheel caused a certain inconvenience in using the tester. The measuring wheel was equipped with a low-pressure tire, as in large flying models. In this way, the authors wanted to reproduce the cooperation of the low-pressure tire with the grass surface of the runway. However, it turned out that a relatively high load on the measuring wheel causes too much deflection of the tire and damage (cracks in the tire). In addition, during the tests to determine the coefficient of surface friction when braking, the tire rotated on the rim of the wheel. Based on the experience gathered during the design and construction of the described tester, the development of a new device, devoid of the described drawbacks, was aimed at.

4.2.2.1.2 Design of the TRT *tester*

The Tire–Runway Tester (TRT) has been designed and developed by the author specially for measurement of grass runway parameters, following the previous tester described earlier. The new design assumptions were as follows:

- measurement of the longitudinal force generated in the wheel-surface contact area;
- based on the measurement, it is possible to determine two components: rolling resistance force and friction force during braking;
- the device includes an indicator of the measured force, calibrated in Newtons, interface for the transmission of measurement data to a portable computer; and
- the tester in form of a portable device, preferably foldable to take up little space (fits in the trunk of a passenger car).

In addition, it was assumed that the force sensor should be simple and cheap to produce, which would facilitate the implementation of the device for practical applications on grass airfields.

In contrast to the earlier design, the wheel tester is based on two, instead of three, roadwheels, on which the surface reacting forces are acting during test runs. Measurement is performed at the so-called "walker speed", that is, 5–8 km/h. This does not correspond to the measurement conditions for the touchdown/liftoff speed of a typical light airplane, but it has been assumed that the manual method cannot be performed. Moreover, as the research shows, the highest values of the rolling resistance coefficient and adhesion occur at minimum speeds. Thus, the measurement result is a reflection of the so-called worst-case scenario.

The wheel–surface force is measured by the means of a simple bending force sensor. The spring element of the sensor is operated by a lever to which the brake caliper is attached at the opposite end.

The brake disc is connected to the axle of the wheels. During braking, the brake caliper is loaded in the direction of rotation of the disc but is blocked by a force sensor. In fact, the braking torque is measured, which is converted into a longitudinal force, taking into account the radius of the wheels. Figure 4.9 shows a cross-section of the sensor with some construction details. The lever with the brake caliper is deliberately mounted on the bearing on the wheel axis, because the bearing holds the greatest moment when moving off. While

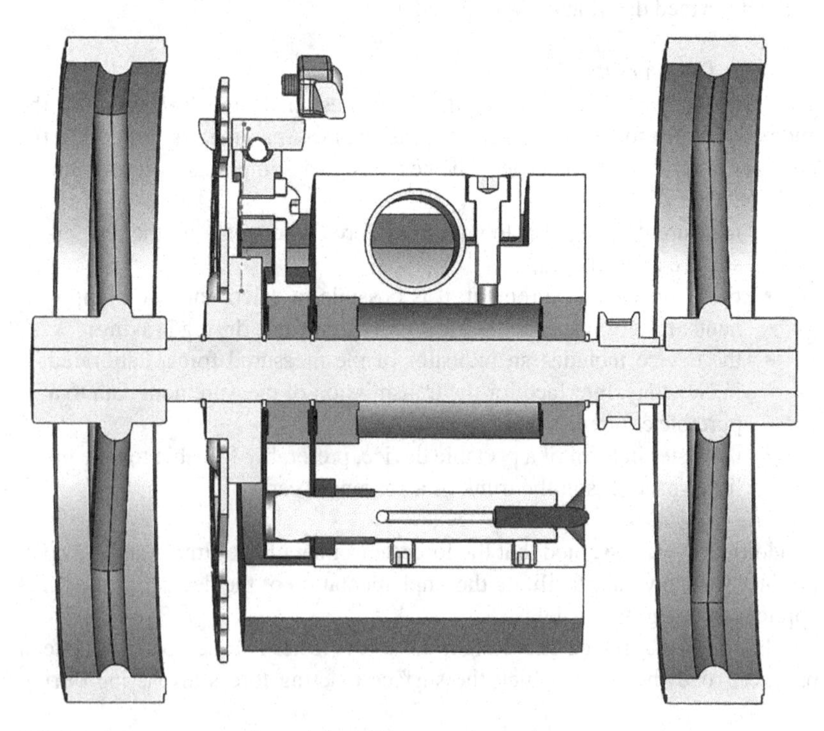

FIGURE 4.9 Cross-section of the TRT-measuring unit.

driving the wheel, the rolling resistance from the bearing will be negligible. The stresses resulting from the cable during braking may have the greatest influence on the force measurement.

Any measurement method of braking force other than the torque measurement method would be sensitive to the inclination of the device while being guided by the operator.

The distance measurement is carried out by an encoder driven by a toothed belt.

4.2.2.1.3 Operation of the tester

The measurement of rolling resistance and braking friction is done by one person. The operation of the TRT tester is simple:

- powering up the device;
- operation mode selection: indication of maximum or minimum values;
- starting data recording;
- rolling resistance measurement: setting the measuring wheel in motion by manually pushing the tester at a speed on the tested surface;
- braking friction measurement: putting the measuring wheel in motion, gradually tightening the brake handle until the wheel is completely blocked – then it slides on the tested surface; the braking friction measurement is recommended by manually pulling the tester backwards; and
- saving the measurement result in the device memory.

Figure 4.10 shows the TRT during trials on a grass airfield.

4.2.2.2 Pull test method[2]

In research praxis, several methods for the measurement of k_{RR} are used. Perhaps one of the simplest is the tow test method. This method is often used for the determination of rolling resistance coefficient in automotive research and testing. A second vehicle tows the test vehicle, and the force needed to pull the test vehicle is measured with the use of a load cell. This method allows the identification of low-speed components of the rolling resistance. In the airfield tests performed by Pytka (2014), a PZL 104 Wilga aircraft and a towing vehicle were attached by a load cell and a cable, positioned so that the cable was taut. With the aircraft fully occupied and fueled, and the brakes released, the towing vehicle pulled the aircraft at a constant speed of about 3–5 km/h for approximately 30 m. The average pulling force divided by wheel

FIGURE 4.10 The TRT tester during tests on a grass runway.

load produces the coefficient of rolling resistance [12]. Figure 4.11 shows a typical setup for pull test method.

The calculation of rolling resistance from experimental data obtained in the pull test is simple, since the method provides time histories of the tow resistance force, which is equal to rolling resistance, as shown in Figure 4.12. The measured values are arithmetically averaged. It is important to use data reduction to choose the right range for averaging: this range should not start from the very beginning of the recorded test but at a moment when the measured values of rolling resistance is stable. In order to determine the rolling resistance for different wheels of the landing gear, it is necessary to determine the distribution of the force on all the aircraft wheels. With the assumption that the aircraft was tested on a level surface, the rolling resistance force distribution over the wheels is proportional to the aircraft weight distribution. To calculate the distribution, the aircraft has to be weighted. In case of the Wilga aircraft, which is a taildragger, the rolling resistance is calculated for the front wheels (superscript F) and for the tail wheel (T) separately, using Eq. (4.11):

$$k_{RR}^{F,T} = \frac{W^{F,T}}{W} \times \frac{F_{RR}^{A}}{W} \tag{4.11}$$

where $W^{F,T}$ is the front or tail wheel load, W is the the total weight of the aircraft, and ${F_{RR}}^{A}$ is the averaged value of rolling resistance from the tow test.

FIGURE 4.11 The experimental setup for a pull test to obtain rolling resistance of airplane landing gear wheels on grass runway (upper photograph) and on a snow-covered runway (lower photograph).

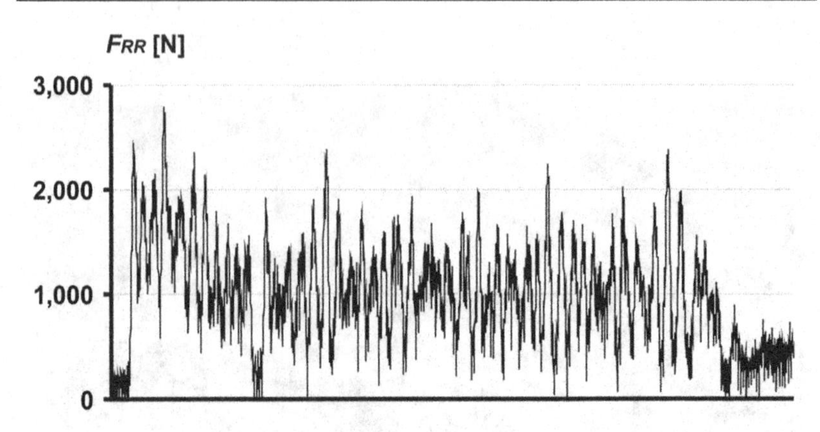

FIGURE 4.12 Sample results of the rolling resistance measurement with pull test method.

Source: Reprinted by permission of the American Institute of Aeronautics and Astronautics, Inc., from [12] (copyright © AIAA)

This method provides a complex value of resistance consisting of drag due to soil or surface deformation and drag due to rolling, both at low speeds. In the cited study, rolling resistance was measured for three different grass surfaces, namely for model airplane airfield (a smooth surface, grass of about 2–3 cm high) and rough grass surface with short and long grass (10 and 20 cm). The determined values of k_{RR} were between 0.032 for the model airplane airfield and 0.075–0.08 for the rough grass surfaces. The difference between the extreme values is pronounced, but, again, the effect of the length of grass seems to be not significant. This corresponds logically with the results obtained in the other presented methods.

4.2.2.3 Instrumented vehicle method[3]

Another is the instrumented vehicle method. In this method, rolling resistance coefficient is determined with the use of an instrumented ground vehicle. The method, therefore, allows to determine the rolling resistance coefficient for the instrumented vehicle's tires, not aircraft tires. Typically, a wheel dynamometer is mounted on the test vehicle together with measuring equipment (data acquisition system, power supply). The wheel dynamometer is a device mounted between a wheel hub of a vehicle and a modified wheel rim. The dynamometer can be built either with strain gage or piezoelectric transducers. The most advanced version of a wheel dynamometer can measure all forces and moments acting on a wheel: vertical F_Z, horizontal F_X, and transversal F_Y forces as well as the moments: M_Z, M_X, M_Y. Pytka (2014) has developed such a measuring device and used it for the measurements of rolling resistance on grass runway. The dynamometer was installed on a front wheel of a SUV. The vehicle was run over the grass surface at speeds of about 3–5 km/h. Higher speeds were not used because of the surface roughness [12].

For the rolling resistance tests, the vehicle was run in rear-wheel-drive mode (without the front drive), so that a resisting moment was the only force acting on the front wheels.

The rolling resistance coefficient is determined with the use of two measures:

- *My* = a moment acting around the transversal axis;
- *Fz* = vertical force.

The k_{RR} is then calculated as given:

$$k_{RR} = \frac{M_Y}{r_d} \times \frac{1}{F_Z} \tag{4.12}$$

where r_d = dynamic radius of the tire.

The value of *My* is calculated by averaging a portion of the measured data. *Fz* is also an average of maximum values determined simultaneously with the *My*. In this method, a complex coefficient of resistance is determined, and this value summarizes drag due to rolling and drag due to wheel and soil deformation. A sample data set for the calculation of k_{RR} is shown in Figure 4.13.

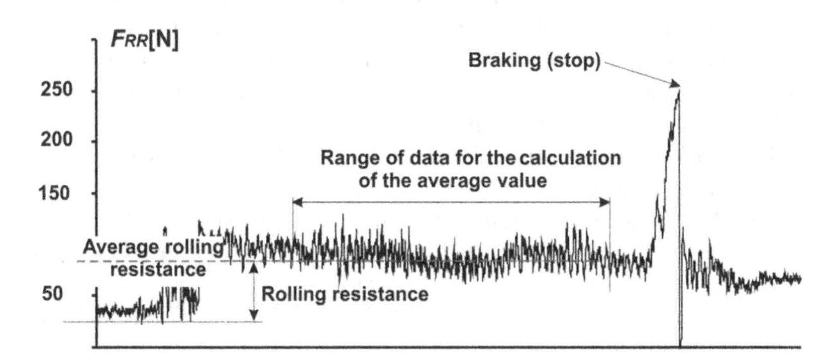

FIGURE 4.13 The calculation of rolling resistance coefficient based on experimental data captured with the use of the instrumented vehicle – wheel dynamometer.

Source: Reprinted by permission of the American Institute of Aeronautics and Astronautics, Inc., from [12] (copyright © AIAA)

4.3 AIRPLANE PERFORMANCE MEASUREMENT METHODS – FLIGHT TESTING

4.3.1 Flight test method for the determination of rolling resistance[4]

The idea of the method is to measure aircraft ground speeds during takeoff runs and, based on this data, to calculate acceleration of the entire airplane then to obtain the rolling resistance coefficient as a result of inverse calculations. For this purpose, on-board instrumentation is required to measure the aircraft's ground speed. Possible measuring instruments are:

- an optical sensor Datron CORREVIT L 400 for ground speed measurements and
- a DGPS multichannel system for measurements and the acquisition of several kinematical measures (velocities, accelerations, etc.).

The CORREVIT is a non-contact sensor that is widely used for ground vehicle speed measurements. Its range of measurement (0–250 km/h) and its resolution (0.1 km/h) make it suitable for the tests. The CORREVIT sensor should be mounted outside the cockpit, on an external rake. For the accuracy and the best sensor function, the sensor must be placed 400 mm above the ground. In a case of an aircraft that changes the height of its center of gravity during the takeoff ground roll, it was practical to use a special version of the CORREVIT sensor, marked 400L, with a tolerance of mounting height of ±130 mm. Electronics, power supply, and a data recorder (DAT Sony) were installed in the cockpit. Figure 4.14 shows the installation of the CORREVIT sensor on a test airplane.

The DGPS system consisted of the sensor module, external antenna, and power supply, and all were installed in the cockpit. The sensor module was connected to a PC notebook. For a better accuracy of the positioning (down to 20 mm), a ground reference station was used during the tests. This station was located on the airfield.

To determine the rolling resistance, speed curves were analyzed for ground speed at the moment of liftoff. The motion resistance was determined by solving a differential equation as given in Eq. (4.13):

$$\frac{dV}{dt} = \frac{g}{W}\left[T - D - k_{RR}\left(W - L\right)\right] \tag{4.13}$$

where V is the aircraft ground speed, W is the aircraft weight, T is the thrust, D is the aerodynamic drag, L is the aerodynamic lift, g is the acceleration due to gravity, 9.81 m/s^2.

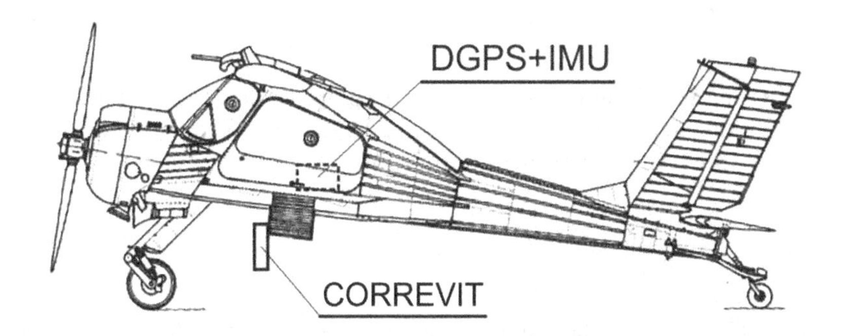

FIGURE 4.14 The installation of the CORREVIT sensor on a test airplane.

This equation includes both aerodynamic forces, lift and drag, as well as the propeller thrust. The aerodynamic forces were calculated with the classic equations. The aerodynamic coefficients were determined from experimental data based on the assumptions:

- in the after-liftoff flare, the aerodynamic lift force is equal to aircraft's weight with respect to climb angle:

$$L = \frac{1}{2}\rho V^2 S C_L = Q\cos\beta \tag{4.14}$$

- the aerodynamic drag is equal to the propeller thrust force minus the longitudinal acceleration of the entire aircraft (determined from the speed data for the after-liftoff flare):

$$D = \frac{1}{2}\rho V^2 S C_D = T - \frac{Q}{g}a_x \tag{4.15}$$

where C_D and C_L are the aerodynamic coefficients, S is the reference area, and ρ is the air density.

The propeller thrust force has been measured: the aircraft was attached to a rigid point by a load cell and a cable, and the thrust was measured at takeoff RPM.

Having measured the true aircraft ground speed and knowing the aerodynamic drag, lift, and thrust forces, the rolling resistance coefficient was calculated using the inverted Euler method. It was assumed that the pitch attitude of the aircraft was constant and consequently the angle of attack and the aerodynamic coefficients C_L and C_D were unchanged during the ground roll. The pilot was asked to perform an "off-three-point" takeoff to assure the aforementioned. Using the Euler method and Eq. (4.13), the takeoff speed history was divided into several small time intervals (i.e., 10), assuming the acceleration was constant during each interval, and the rolling resistance coefficients were obtained for the end of each time interval. Figure 4.15 shows a calculation method based on the experimental data of ground speed time course.

It should be noted that the k_{RR} determined in this method is a summary complex value, containing all components: drag due to high speed, drag due to soil and tire deformation, and drag due to rolling.

The flight test method is methodologically complex. The limitation is the ability of the aircraft to fly from an investigated surface. For example, we were

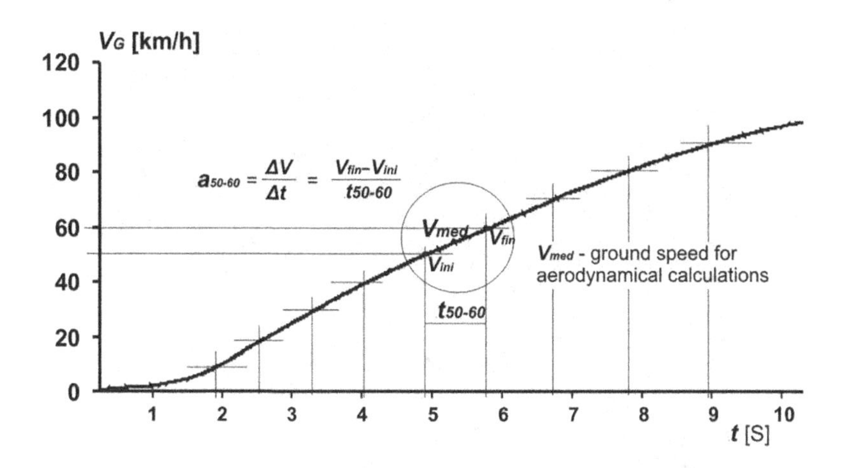

FIGURE 4.15 The calculation of airplane acceleration from the measured ground speed.

Source: Reprinted by permission of the American Institute of Aeronautics and Astronautics, Inc., from [12] (copyright © AIAA)

not able to perform flight tests on the loess soil because the pilots didn't agree due to safety matters. This method can also suffer from some other effects, such as wind or piloting technique.

4.3.2 Flight test method for the determination of ground reaction under landing gear wheel loading[5]

Pytka et al. (2004) have measured the soil reaction against the load of a wheel at the moment of touchdown in terms of soil stresses [13]. Soil stress state was determined with the use of a Stress State Transducer (SST). The device consists of six transducers that measure soil pressures, which are needed for the determination of stress state components: σ_1 – major stress, σ_2 and σ_3 – minor stresses, σ_{OCT} and τ_{OCT} – invariant in the octahedral stress system, normal and shear, respectively [14, 15]. The SST used by the author, measured Ø70 × 50, diameter × height, also included an accelerometer and was installed at 15 cm depth by the simple excavation of the soil (see Figure 4.16). After the SST was emplaced, the soil was backfilled, and the grass was replaced on the surface. The soil over the SST was compacted to

FIGURE 4.16 Emplacing the SST at a depth in grass runway soil substrate.

obtain the initial bulk density. Then, the transducer remained in place for all landing tests.

In the gravity center of the test airplane, the PZL 104 Wilga 35A, a three-axis (X-Y-Z) accelerometer, was installed to collect aircraft acceleration at the moment of touchdown. To ensure that the results are repeatable, numerous trials were needed. Therefore, the tests were performed with the help of four champion pilots, who trained their skills of precision landing. Each pilot performed conventional approach and landings (5% gradient of glide slope, 2.5 m/s descent rate, approach speed 100 km/h, power setting 25–30%) and emergency, zero-thrust landings from the altitude of 300 m above the airstrip, with idle power. More than a 100 landings were performed, and 32 were successful – the left wheel of the aircraft touched the ground at the point above the SST, but only some of them provided satisfactory soil pressure data. In Figure 4.17, the test airplane, PZL 104 Wilga 35A, is visible at the moment of touchdown, during the test flying.

Soil stress state components were calculated from the soil pressure data obtained by the SST. For further analysis, only those tests in which significant values of stress occurred were chosen. Detailed numerical data of the soil stresses can be found in the reference paper [13]. To recognize if the wheel touched the surface exactly over the SST, the relation of peak values of major stress σ_1 and the acceleration was analyzed. If the peak acceleration appeared first, it meant that the wheel touched the surface in the sensitive area of the accelerometer before the SST. The area of sensitivity for the SST stress sensors is smaller, so if both peak σ_1 and acceleration appeared simultaneously, the point of touching the surface by the wheel could be estimated more precisely.

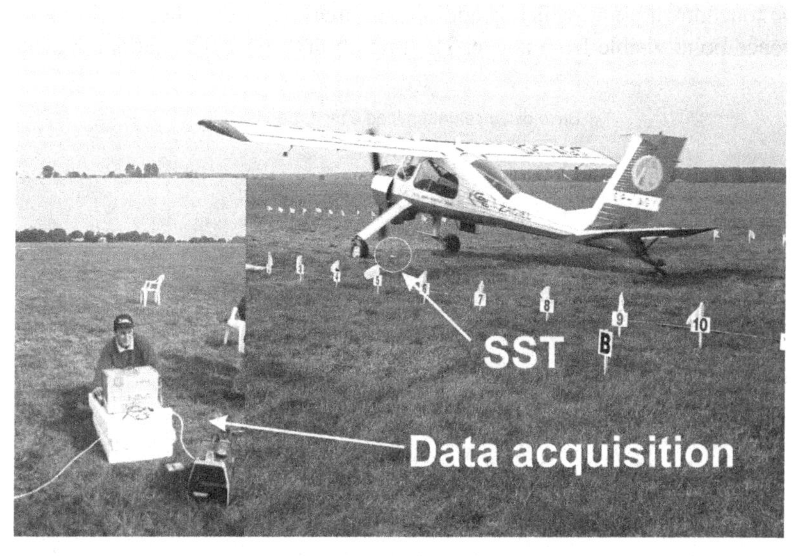

FIGURE 4.17 The PZL 104 Wilga 35A touches down the grass runway. The SST is in soil beneath the left wheel.

4.3.3 Ground observer method for takeoff and landing ground distance measurement

This is probably the simplest method of measuring takeoff or landing distance that can be carried out in the field. It basically determines the liftoff or touchdown point relative to markers along the runway. The ground observer observes the takeoff/landing from a safe distance, which in the case of GA airplanes with piston drive is approximately 15–20 m (counting from the edge of the runway). Markers in the form of plates with a distance in meters are placed along the edge of the strip. A typical configuration of markers is shown in Figure 4.18.

During the takeoff measurement, the airplane starts the run-up from a predetermined point in relation to which the positions of individual markers are determined. The "zero" point is established experimentally on the basis of several (two to three) test starts. During the measurement, the observer is near the "zero" point. Visually, he or she marks the point where the wheels are separated from the road surface relative to the nearest marker and records the distance. The result is obtained by subtracting or adding the length determined by the observer in relation to the base starting point – the zero point is marked.

The landing measurement requires two observers. The first one is near the touchdown point, which is additionally marked so that the pilot has a reference point visible from above. The first observer visually marks the actual

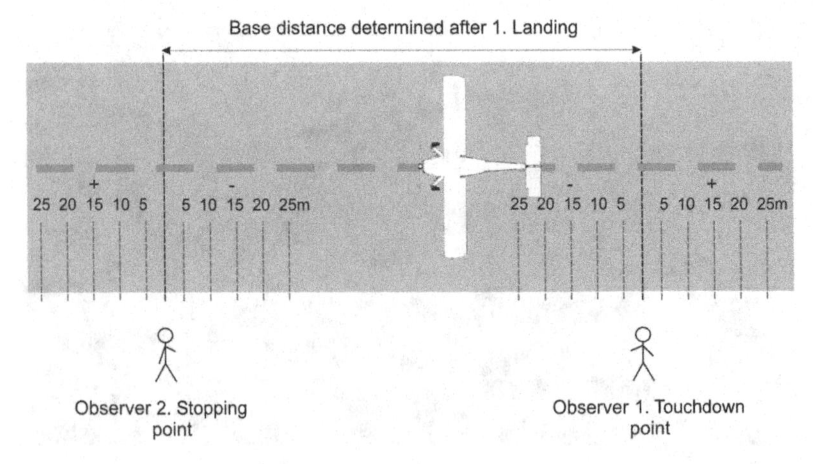

FIGURE 4.18 A schematic of the markers used as reference for the ground observers.

touchdown point with respect to the markers that are positioned in front of and behind the line. On the other hand, the second observer marks the place of stopping the airplane with a marker placed at the point where the wheels of the main landing gear stopped. It is necessary to measure the distance between the touchdown point and the stopping point.

The accuracy of this method depends on the adopted resolution of the marker's position, with the optimal resolution of 5 m. At higher resolutions, for example, 2 m or 1 m, there are problems in determining the liftoff or touchdown point visually. In the case of typical piston airplanes, for which the length of the takeoff or rollout run is approximately 200–250 m, the measurement uncertainty is below 2%. The method is simple and precise. The disadvantage is the need to involve people as ground observers.

4.3.4 Video camera method

One version of this method is based on the use of a camera and a sighting bar, placed between the camera and the runway. This method is allowed by the FAA for measuring takeoff and landing distances, according to the Advisory Circular 23–8A. Another method uses various systems detecting the touchdown point of the wheels, for example, a system of transversely placed steel strings connected with acceleration sensors. This version of the measurement method is used in precision-landing competitions.

4.3.5 Takeoff and landing measurements with the use of AI sensor[6]

4.3.5.1 Introduction

For determining the ground performance of light airplanes operating from unsurfaced airfields and opportune grass landing sites, measuring the actual takeoff or rollout distance is of great importance. Often, when a pilot arrives at a given landing site, he is not able to determine with sufficient certainty whether the takeoff on the way back will be safe, especially if the length of the available runway is small, and the condition of the turf raises additional doubts (long, tall grass, moist, soft soil). The methods of measuring ground performance used in industrial practice are inadequate to the aforementioned conditions, mainly due to the difficulties in implementation, the need to use specialized equipment, and the dependence on technicians making measurements on the ground.

Rather, it is preferable to use a method based on a simple, lightweight on-board unit. The author has made an attempt to use the INS/GNSS (Inertial Navigation System/Global Navigation Satellite System) sensor-based method to determine the ground distances of a PZL 104 Wilga 35A plane, taking off and then landing at the airport with a grass strip. The results of measurements with the use of a RTK (Real Time Kinematic) device were promising; however, the method suffered from the inconvenience of the need to manually analyze the acceleration of the center of gravity of the aircraft. In addition, professional measuring instruments were used, which were difficult to access, expensive, and of relatively large sizes [16, 17].

Artificial Intelligence and neural networks offer great opportunities for the recognition of flight phases, which is the basis of the method based on acceleration measurements. In the problem of recognizing aircraft flight phases, we deal with acceleration time courses, which are of quasi-periodic character. The form and size of the courses are dependent upon flight condition, flight phase, and the actual attitude of airplane. The time courses of acceleration may contain essential information about the corresponding flight configuration, propulsion system function, etc.

Among different deep learning methods, Convolution Neural Networks (CNNs) have properties that make them very suitable for recognizing the movement of moving objects. Ignatov in 2018 drew attention to the particularly beneficial effects of using CNN to recognize human activity [18, 19]. Based on this idea, the author attempted to use convolutional networks to identify phases (states) of aircraft flight [20–22].

Taking into account the aforementioned studies and considerations, it was aimed to develop and test a new sensor, on the basis of IMU/GNSS measurement platform, utilizing a CNN for identifying of characteristic phases of flight. The prototype sensor, primarily called IMUMETER, was installed on a test airplane, and a series of test flight has been aimed in order to examine the proposed method.

This section of the book was created on the basis of research paper “Measurement of aircraft ground roll distance during takeoff and landing on a grass runway” by Jarosław Pytka, Piotr Budzyński, Paweł Tomiło, Joanna Michałowska, Dariusz Błażejczak, Ernest Gnapowski, Jan Pytka, Kinga Gierczak published in *Measurement*, Vol. 195, 2022, p. 111130, Copyright 2022, with permission from Elsevier.

4.3.5.2 Flight phases’ recognition by means of neural network

The landing of the aircraft can be divided into five characteristic phases: approach, flare, touchdown, rollout, and stop. To correctly determine the rollout distance, it is necessary to know the point of touchdown and the point of stop (end of the rollout). Similarly, the takeoff of an airplane can be divided into characteristic phases, which are described by measurable quantities, such as vertical acceleration of the center of gravity of the airplane, magnetic field vector, and altitude. The measurement of vertical acceleration of the entire airframe can be used to detect landmarks during landing, and this was inspired by methods of human activity recognition using accelerometers [18]. By analyzing the course of this parameter, the touchdown point and the end of the rollout can be recognized. While this has been performed manually in the past [23], the present version of the method assumes automation that is possible using CNN method.

4.3.5.3 Neural network development

Initially, two separate neural networks were developed, one for takeoff and another for landing. Both models were based on characteristic points or phases of flight and required separate training. These two models have been experimentally validated as part of the flight test campaign (more details in the next section). It has been found that, from a practical point of view, it will be more advantageous to have one CNN-based model capable of describing and recognizing the phases of flight, both during takeoff and landing. Another network was developed for five classes, which correspond to five different phases of aircraft flight.

The artificial neural network uses eight input values; accelerometer and magnetometer data in x, y, and z axes; and altitude and speed of an airplane from 1 min, so the input size is $60 \times 8 \times 1$. In order to recognize both local and global features, the input is separated into two branches. In the first one, input data is fed to the convolution layer with 16 filters and kernel size 3×4 with stride 1×1 and then to max pooling layer, kernel size and stride 3×2. In the second branch, data is fed to the convolution layer similar to the first branch, then to max pooling layer with kernel size and stride 1×2. The second step in this layer is to put data to the convolution layer with 16 filters and kernel size 3×2 and stride 1×1. To achieve the same size in both branches, it is necessary to use max pooling layer in the second branch with kernel size and stride 3×1. Both branches are concatenated (size: $20 \times 4 \times 16$), and the data goes to convolution layer – 32 filters, kernel size 2×2, and stride 1×1, and then data goes to max pooling layer with kernel size and stride 4×4, so the size of the data is $5 \times 2 \times 32$. Since the data flows through the two parallel network's branches, it was named "double – branch CNN" (DB_CNN) in contrast to the older neural network, proposed in [19, 21], in which the data flowed through only one branch ("single-branch" CNN, SB_CNN). Before data is fed to the fully connected layer, it is flattened. First, it goes to the fully connected layer with

64 neurons. For training purposes, to reduce over-fitting, there a dropout layer is also added before last dense layer. In the training process, we have used a dropout layer with 0.2 rate of turning off neurons, because the next dense layer has the same number of neurons as the previous one. The third dense layer has 32 neurons. All the fully connected layers' activation function is Rectified Linear Unit (ReLU). The last layer has five neurons, and it is activated by softmax function. For the network parameters optimization, the Adam algorithm was used with the parameters: $\alpha = 0.001, \beta_1 = 0.9, \beta_2 = 0.999, \epsilon = 10^{-8}$. The size of the DB_CNN neural network model. h5 file is 389 kB, and that of the SB-CNN model file is 563 kB.

One important feature of the new neural network is that it includes two filters that focus on collecting data series from both the total input data and its smaller parts for information about the occurrence of a peak at the time of touchdown and the change of acceleration at the time of takeoff.

The algorithm of the neural network model uses the following data as input: the acceleration of the airframe in three mutually perpendicular directions, magnetic field vector, and data from the GPS sensor: aircraft ground speed and altitude. All data necessary to train and run the CNN model was measured at a sampling rate of 23 Hz.

The DB_CNN neural network has been trained using data collected during flight tests performed with two airplanes: a Morane Saulnier MS880 and a Cessna 172. The MS800 is a General Aviation (GA) aircraft, powered by a 162-kW piston engine, and it has an undercarriage with a nose wheel and is a low-wing airplane. The Cessna 172 aircraft is powered by a 134-kW engine. It has a non-retractable landing gear with a nose wheel and is a high-wing airplane.

Both planes have a cabin with four places for the crew and passengers. The choice of airplanes was intentional, as the purpose was also to test whether the method is sensitive to the airframe structure: low wing versus high wing. For a low-wing plane, the center of gravity is usually above the wings, and for a high-wing plane, the opposite is true. This influences on the character of the vibrations excited during the ground roll.

4.3.5.4 Instrumentation

The Raspberry PI 4B single-board, mini-computer was used as the hardware platform of the IMUMETER for measurements. The Raspberry PI OS operating system has been installed on the device. The neural network model was run and trained with the use of tensorflow library in python language.

The second modification to the measurement instrumentation was the use of sensors with higher parameters. The following sensors were used:

- GPS RTK sensor of 25-Hz navigation rate with a declared horizontal position accuracy of 2.5 m and 0.01 m (with and without RTK, respectively); and
- accelerometers, with configurable measuring range ±2/±4/±8/±16 g, refresh rate up to 952 Hz; and
- magnetometers, with configurable measuring range of ±4/±8/±12/±16 gauss, refresh rate up to 80 Hz; and
- altitude and velocity sensors.

The aforementioned sensors have been integrated with a single-chip computer. The complete measuring device, IMUMETER, additionally had a display that allowed the control of the operation of the measuring system and a power supply battery. The main unit, including the computer and sensors, was installed at approximately the center of gravity of the airplane (see Figure 4.19).

In addition, a base station was built consisting of the following devices:

- antenna which is capable of receiving the L1/L2 bands for GPS, GLONASS, Galileo, and BeiDou constellations;
- GNSS receiver (GPS, GLONASS, Galileon, BeiDou); and
- server (PC).

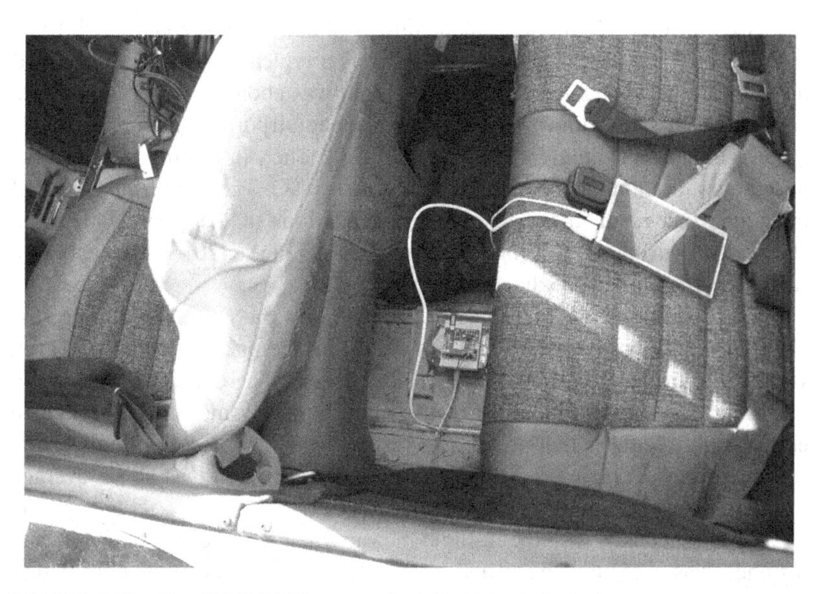

FIGURE 4.19 The IMUMETER sensor installed in a test airplane.

The base station was located in a house approximately 5 km from the airport where flight tests were conducted. The coordinates of the base station position were as follows: latitude 51.237063° and longitude 22.601365°. Communication between the IMUMETER device installed on the test plane and the base station was made via the Internet using the NTRIP server. The diagram of the entire measurement system is shown in Figure 4.20.

4.3.5.5 Calibration of the method

Dataset for training of the neural network was collected by performing a minimum of 10 flights for takeoff and landing for each of the two airplanes. The results of learning the neural network are included in the cited reference [22]; nevertheless, it was found that these results are very good, similar or even better compared to those reported in [21]. It was also found that one network fits two different airplanes, while teaching the network to a given airplane requires data from measurements performed on a given object.

The flight test campaign was performed with the use of the MS880 and the C172 airplanes. Two pilots took part in the flight test program, and both had similar ratings and practical experience. The aircrafts were flown with two persons onboard – one pilot and one instrumentation engineer who maintained the computer.

A total of 109 flights were made during about 8 flying hours. Each time, the measurement flight looked similar, it was a typical four-turn pattern, and the typical altitude achieved during the flight was about 200 m. Flights were performed on grass runway 29 of the Radawiec airport, near Lublin, southeastern Poland, in the months of August to October, in the morning or evening hours, at an air temperature of about 10–18°C. The pilot used a typical technique that resulted from the recommendations of the Pilot's Operating Handbook. During the landing rollout, in the final phase (20–30 meters before stopping), the pilot used wheel brakes.

The conditions on the runway were as follows: grass surface and grass blades' height approximately 7–12 cm. The surface of the runway was even with slight unevenness, typical for a grass airfield. There are two runways at the Radawiec airport, and both of them are regularly mowed and rolled to eliminate unevenness that arises as a result of the activity of animals or the tufting of grass. During the flight test period, vegetation was above-average lush due to heavy rainfall in the summer of 2021.

Simultaneously, with the in-flight measurements, the length of the ground distance was measured by visual observation. The measurement resolution of the reference method for measuring the takeoff and landing distance was 2.5 m, and the measurement error was about 1.25%. The method is based on

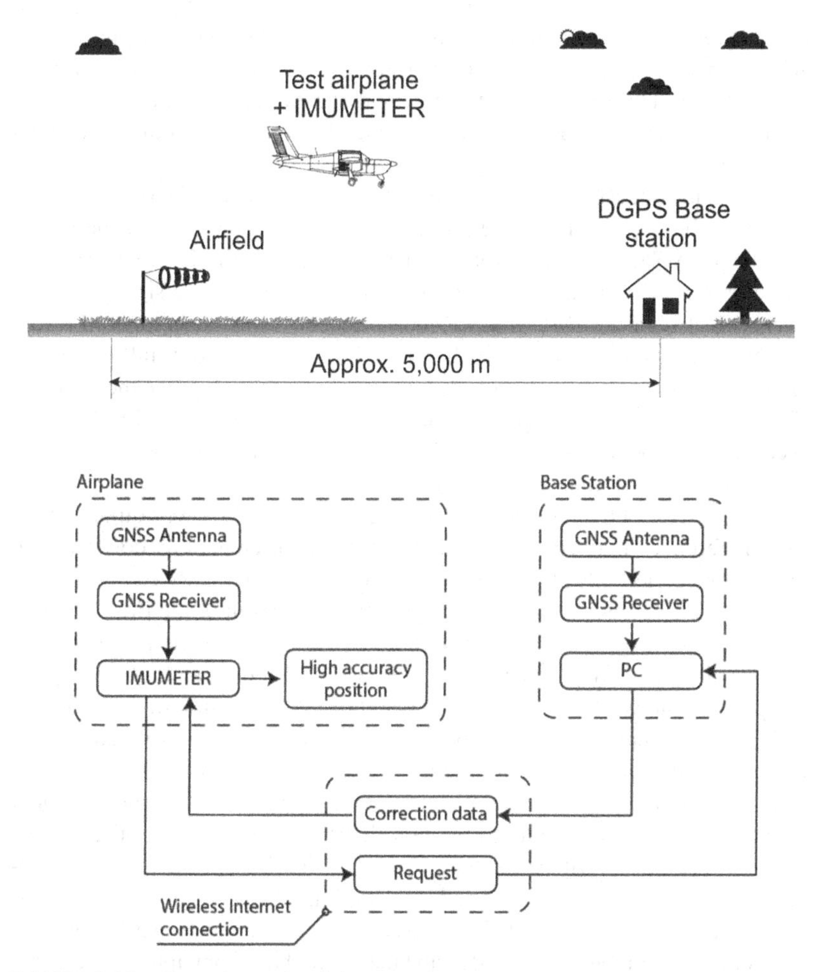

FIGURE 4.20 The schematic of the complete measurement system.

Source: Reprinted from "Measurement of aircraft ground roll distance during takeoff and landing on a grass runway", by Jarosław Pytka, Piotr Budzyński, Paweł Tomiło, Joanna Michałowska, Dariusz Błażejczak, Ernest Gnapowski, Jan Pytka, Kinga Gierczak published in *Measurement*, Vol. 195, 2022, p. 111130 (copyright 2022 with permission from Elsevier)

another method that has been used for years in sports aviation during landing accuracy competition.

Results of measurements of ground roll distances, carried out with the use of the presented method (which is IMUMETER method), were compared with the results of measurements using the reference method. Each individual measurement was analyzed, comparing the IMUMETER measurement with the reference method for every flight, and the average difference between the measurements of the tested device and the reference method was also calculated.

First, the measurement was carried out with the IMUMETER device placed on the floor lining of the aircraft cabin. The mean value of the difference of the MS880's landing rollout distance between the IMUMETER and the reference measurement was 8.09%. In several cases, quite significant differences between the IMUMETER-measurement results as compared to the reference method were observed; therefore, it was decided to install the device directly on the metal elements of the aircraft structure (but still in the cockpit). The measurements were repeated, and their results compared with the reference method. The obtained mean difference was 2.92%, which was considered a much better result than assumed (7%).

The results of the measurements of the takeoff roll showed quite significant differences between the IMUMETER-measurements as compared to the reference method (by an average of 28%), and practically each time, the result of the reference method was higher than the result of the IMUMETER measurement. Repeating the measurements did not solve the problem. The average absolute difference was approximately 35m, that, at the liftoff speed of approximately 55 mph = 25 m/s, results in a delay time of 1.4 s, which is much more than the typical observer's reaction time (approximately 1 s). It was found that a systematic error in the reference method is possible, consisting in the erroneous observation of the lifting-off of the landing gear wheels from the ground due to the fact that the landing gear of the MS880 aircraft is equipped with telescopic shock absorbers, which prolong the contact of the wheels with the ground. In addition, the fact that the wheel just after liftoff, for a moment, still touches the grass for a while, while optically still remaining in contact with the surface, could have contributed to erroneous observations with the reference method. Here, it should be emphasized that while the reference method is used in precision flying contests, it has not yet been sufficiently tested for takeoff. Therefore, it was concluded that the measurement with the IMUMETER method can be considered correct. It is justified, taking into account the IMUMETER-measured average takeoff roll distance, which was 140 m, that is, differed by 7% from the takeoff ground roll according to the POH data for the MS 880 aircraft, which is 130 m.

However, due to the aforementioned doubts, it was decided to repeat the measurement with the aircraft equipped with the main landing gear without telescopes. This was the second aircraft used in this study – the Cessna 172. Of course, this was before ground roll measurement flights and the measurements of acceleration, magnetic field, and GPS signals were performed to train the neural network for this aircraft. Then, the appropriate measurements of the run-up length were carried out using the IMUMETER device method and the reference method at the same time. A much better correlation of the measurement results for both methods was observed, and the average difference was 6.31%, which was considered a satisfactory result. For the sake of accuracy, it should be mentioned that also in the case of the C172, the undercarriage of which does not have shock absorbers (the function of the wheel suspension is performed by a spring landing gear), higher results of the reference method were observed. Thus, subjective observations, assisted by the markers alongside the runway, suffer from an error due to the observer's eye response and the inability to distinguish the liftoff point when the "optical" wheel of the landing gear is still at the surface. Additionally, the observation conditions could be made worse by the presence of grass in the strip.

The achievement of the DB_CNN network performance result was influenced by the use of two branches in the network structure as well as the addition of input data: magnetic field vector components, altitude, and airplane speed. The primary input data of vertical and horizontal acceleration are very expressive in terms of identifying the liftoff and touchdown points, while the added input data significantly contributed to the recognition of individual flight phases, which influenced the final result in terms of precision and accuracy of recognition.

The second important feature of the developed DB_CNN model is the small size of the model, namely 389 kB, while for the SB_CNN model, it is 563 kB. This is an important property for the practical use of the model – for example, when using low-cost single-board computers or smartphones. The improvement resulted from the modification of the network structure and is significant despite the addition of more data series. It is also a significant advantage taking into account the size of typical network models based on computer vision, which can be several megabytes.

Occasional false recognition results, and erroneous takeoff measurements or missing data may result from the fact that the calculations were performed in real time. Neural network prediction results and GPS data are collected asynchronously and can cause an error bacause of a limited computing power of the hardware platform. This problem was reported and analyzed by Petritoli et al. (2021); to solve it, either a new, more efficient computer, adapted to AI-applications, should be used or further modification of the neural network should be performed, for example, with attention layers. One good solution could also be to use a 11 DOF (*Degree of Freedom*) inertial measurement unit to acquire more data for the neural network. The next solution to reduce error would be to perform the segmentation of flight data time series after the completion of a flight, which would allow the interpolation of GPS data to match the neural network prediction time.

When discussing the practical course and the results of takeoff and landing measurements, it can be seen that compared to the methods proposed by Bakunowicz and Rzucidło [24], the presented method is, above all, easier to use and more accessible, due to the lack of the need to use complicated ground equipment and additional ground personnel. Additionally, the result is obtained immediately, without the need to post-process the results. We have obtained measurement accuracy of 2.92 % and 6.31 % in terms of average difference between the IMUMETER and the reference method. This result is satisfactory having in mind the use of the new method in flight test practice.

4.3.5.6 Conclusion

The developed method of measuring the takeoff and landing distance of an airplane is simple to use and, at the same time, sufficiently accurate to be used in this study. One person is needed to operate the IMUMETER on-board unit, and the results are available in real time, also on any ground device. In the next chapter, the results of the measurements of the takeoff and rollout length obtained using the method described before will be presented.

4.3.6 Method for airplane stability during takeoff and landing ground roll

Pytka et al. (2013) carried out tests of the stability of the aircraft's movement during the takeoff ground roll. It was assumed that the experiment will be the basis for determining the stability of an aircraft with a three-wheel landing gear system. On the basis of the data collected during the flight tests, the system identification procedure was carried out, a mathematical model was derived, and the characteristic equation was determined. The analysis of the location of the roots of the characteristic equation allowed for the assessment of the stability of the aircraft motion (Pytka 2013).

4.3.6.1 The test airplane and the method

The PZL 104 Wilga 35A aircraft was used in the tests. The preparation for the tests was done by installing the measuring equipment, which included a DGPS receiver block with the RT3002 inertial platform. The device was placed near the center of gravity of the plane, and, in addition to that, on board there was a 12 V/50 AH power supply battery, a receiving antenna, and a portable computer for controlling the measurement and data recording. The DGPS ground reference station remained outside the aircraft.

4.3.6.2 *Flight tests' campaign*

The flight tests consisted of multiple takeoffs and landings of the airplane on a grass surface of the Radawiec airfield, near Lublin. In order to investigate the effect of grass height, the tests were carried out twice in the early spring (April), when the grass at the airport is short (approx. 5–10 cm), and in late spring (end of May), when the grass is relatively high, approximately of height 20–25 cm, and this is the time of the first mowing.

The plane flew with a crew of four: the pilot and test coordinator in the front seats, test engineer operating the measuring system, and an assistant in the rear seats. For both cases of grass height, 10 flights were performed in order to obtain representative data.

4.4 CONCLUSION

In this chapter, methods for measuring and analyzing the performance of an airplane on a grass runway have been presented. Also, methods for measuring parameters that characterize the surface of a grass runway have been described. Some of the methods originate from terramechanics, for example, the cone penetrometer and the TDR moisture meter. Other methods discussed here rely on measuring devices and systems specifically developed for this study, such as a wheel dynamometer and artificial intelligence IMUMETER sensor. The information and data contained in this chapter may encourage to make attempts, both in terms of following the author's projects as well as to design and develop new sensors and measurement systems. The main goal of this chapter was to develop measurement methods necessary for the verification and parameterization of the models presented in the next chapter. The devices and systems as well as the procedures included in these methods have been successfully tested in field tests at a grass airfield.

NOTES

1 Reprinted from"Wheel dynamometer system for aircraft landing gear testing", authors: Pytka J., Józwik J., Budzyński P., Łyszczyk T., Tofil A., Gnapowski E. and Laskowski J., published in: *Measurement* (148) 2019, p. 106918. Copyright 2019 with permission from Elsevier.

2 Reprinted by permission of the American Institute of Aeronautics and Astronautics, Inc., from [12]. Copyright © AIAA.
3 Reprinted by permission of the American Institute of Aeronautics and Astronautics, Inc., from [12]. Copyright © AIAA.
4 Reprinted by permission of the American Institute of Aeronautics and Astronautics, Inc., from [12]. Copyright © AIAA.
5 This chapter was published in *Journal of Terramechanics*, Vol 40, Pytka J., Tarkowski, P., Dąbrowski, J., Bartler, S., Kalinowski, M., Konstankiewicz, K. Soil stress and deformation determination under a landing airplane on an unsurfaced airfield, pages 255–269. Copyright Elsevier and International Society for Terrain – Vehicle Systems, 2004 [13].
6 Reprinted from "Measurement of aircraft ground roll distance during takeoff and landing on a grass runway", by Jarosław Pytka, Piotr Budzyński, Paweł Tomiło, Joanna Michałowska, Dariusz Błażejczak, Ernest Gnapowski, Jan Pytka, Kinga Gierczak, published in *Measurement*, Vol. 195, 2022, p. 111130. Copyright 2022 with permission from Elsevier.

REFERENCES

1. U.S. Army and Air Force. 1994a. *Planning and design of roads, airfields, and heliports in the theater of operations – airfield and heliport design*, Vol. I. U.S. Army FM 5-430-00-1/Air Force AFJP 32–8013. Washington, DC: U.S. Army and Air Force
2. U.S. Army and Air Force. 1994b. *Planning and design of roads, airfields, and heliports in the theater of operations – airfield and heliport design*, Vol. II. U.S. Army FM 5-430-00-2/Air Force AFJP 32–8013. Washington, DC: U.S. Army and Air Force
3. Mc Allister M. A rig for measuring the forces on a towed wheel. *Journal of Agricultural Engineering Research*, 24(3), September 1979, pp. 259–265
4. Wismer R.D., Luth H.J. Off-road traction prediction for wheeled vehicles. *Journal of Terramechanics*, (10), 1973, pp. 49–61
5. Shoop S.A., Diemand D., Wieder W.L., Mason G., Seman P.M. Predicting California bearing ratio from trafficability cone index values. Technical Report ERDC/CRREL TR-08–17 October 2008
6. Shoop S.A., Coutermarsh B., Cary T., Howard H. Quantifying vegetation biomass impacts on vehicle mobility. *Journal of Terramechanics*, 61, 2015, pp. 63–76
7. Wieder W.L., Shoop S.A. State of the knowledge of vegetation impact on soil strength and trafficability. *Journal of Terramechanics*, 78, 2018, pp. 1–14
8. Kuchler M., Schrupp R. *Mehrkomponenten-motorradmessnabe/multiaxial motorcycle wheel load transducer*. Dusseldorf: VDI-FVT Jahrbuch, VDI Verlag, 2002, pp. 91–119 (In German with English summary)
9. Loh R., Nohl F.W. Mehrkomponenten-Radmessnabe. Einsatzmoeglichkeiten und Ergebnisse. (in German with English summary). *ATZ (Automobiltechnische Zeitschrift)*, 94(1), 1992, pp. 44–53

10. Pacejka H.B., Wallentowitz H. *Tyre and vehicle dynamics*. Elsevier Butterworth – Heinemann, Oxford, UK, 2004
11. Pytka J. Budzynski P., Tarkowski P., Piaskowski M. A portable wheel tester for tyre-road friction and rolling resistance determination. *IOP Conference Series: Materials Science and Engineering*, 2016. doi:10.1088/1757-899X/148/1/012025
12. Pytka J.A. Identification of rolling resistance coefficients for aircraft tires on unsurfaced airfields. *Journal of Aircraft, AIAA Journal of Aircraft*, 51(2), 2014, pp. 353–360
13. Pytka J., Tarkowski P., Dąbrowski J., Bartler S., Kalinowski M., Konstankiewicz K. Soil stress and deformation determination under a landing airplane on an unsurfaced airfield. *Journal of Terramechanics*, 40, 2004, pp. 255–269
14. Nichols T.A., Bailey A.C. Johnson C.E., Grisso R.D. A stress state transducer for soil. *Trans ASAE*, 30(5), 1987, pp. 1237–1241
15. Way T.R., Johnson C.E., Bailey A.C., Raper R.L., Burt E.C. Soil stress state orientation beneath a tire at various loads and inflation pressures. *Journal of Terramechanics*, 33(4), 1996, pp. 185–194.
16. Pytka J., Budzyński P., Józwik J., Łyszczyk T., Gnapowski E., Laskowski J. Measurement of takeoff and landing ground roll of airplane on grassy runway. In *Proceedings of the 6th International Workshop on Metrology for Aerospace (MetroAeroSpace)*, Torino, Italy, 19–21 June 2019
17. Pytka J., Budzyński P., Józwik J., Michałowska J., Tofil A., Łyszczyk T., Błażejczak D. Application of GNSS/INS and an optical sensor for determining airplane takeoff and landing performance on a grassy airfield. *Sensors*, 19, 2019, p. 5492. doi:10.3390/s19245492
18. Ignatov A. Real-time human activity recognition from accelerometer data using Convolutional Neural 474 Networks. *Applied Soft Computing*, 62, 2018, pp. 915–922, 475
19. Tomiło P. Metoda pomiaru długości startu i lądowania samolotu z wykorzystaniem sztucznych sieci neuronowych (A method of measuring the take-off and landing distance of an airplane with the use of artificial neural networks), PhD thesis, in preparation, Lublin University of Technology
20. Pytka J., Budzyński P., Tomiło P., Michałowska J., Gnapowski E., Błażejczak D., Łukaszewicz A. IMUMETER – AI-based sensor for airplane motion measurements. *IEEE International Workshop on Metrology for AeroSpace, MetroAeroSpace 2021 – Proceedings*, 2021, pp. 692–697, 9511767
21. Pytka J., Budzyński P., Tomiło P., Michałowska J., Gnapowski E., Błażejczak D., Łukaszewicz A. IMUMETER – A convolution neural network-based sensor for measurement of aircraft ground performance. *Sensors*, 21, 2021, p. 4726
22. Pytka J., Budzyński P., Tomiło P., Michałowska J., Błążejczak D., Gnapowski E., Pytka J.D., Gierczak K. Measurement of aircraft ground roll distance during takeoff and landing on a grass runway. *Measurement*, 195, May 2022, p. 111130
23. Pytka J., Tarkowski P., Dąbrowski J., Zając M., Konstankiewicz K., Karczewski L. Determining the ground roll distance of an aircraft on unsurfaced airfield. *Proceedings of the 10th European Conference of the ISTVS*, Budapest, Hungary, October 2006

24. Bakunowicz J. Rzucidło P. Detection of aircraft touchdown using longitudinal acceleration and continuous wavelet transformation. *Sensors*, 20, 2020, p. 7231. doi:10.3390/s20247231
25. Pytka J., Józwik J., Budzyński P., Łyszczyk T., Tofil A., Gnapowski E., Laskowski J. Wheel dynamometer system for aircraft landing gear testing. *Measurement*, (148), 2019, p. 106918

Results

5

5.1 GRASS RUNWAY CHARACTERIZATION

The research was carried out at two grass airfields, namely:

- Radawiec airport, near Lublin; and
- Depułtycze Królewskie airport, near Chełm.

Both airports are located in south-eastern Poland. Radawiec airfield has a loess-sandy soil. Grass turf is a mixture of various grasses, fescue, ryegrass, panacea, and burrow. Additionally, there is low meadow vegetation. Overall, the vegetation at this airport has remained unchanged for almost a 100 years from the time when the Aero Club of Lublin was established in Radawiec in 1927. The runway of the Radawiec airport is rolled once a year in order to compensate for unevenness, and the grass is mowed two or three times a year. On the runway, the grass is partially degenerated, and the wheel marks along the runway are visible (see Figure 5.1). In the rest of the area, except for the runway, the vegetation is lush and mowed less frequently. It was possible to use this part of the airport to conduct flight tests on grass up to approximately 25-cm high.

At the Radawiec airfield, there is a separate runway for model airplanes. The surface of this runway is covered with grass that grows very densely, and its height does not exceed 3 cm. Moreover, the unevenness of the surface is much smaller compared to the main runway. The runway for model airplanes has a width of 70 m and a length of about 250 m, thanks to which it was possible to carry out flight tests with a STOL plane, such as the PZL 104 Wilga 35A.

Depułtycze Królewskie airport is built on marly soil. The sod was planted 20 years ago and is periodically renovated and fertilized. The runway is overgrown with grass of a homogeneous composition, with the predominant species of field grass, resistant to creasing. There are no other plant species. The grass is mowed twice a year.

DOI: 10.1201/9781003312765-5

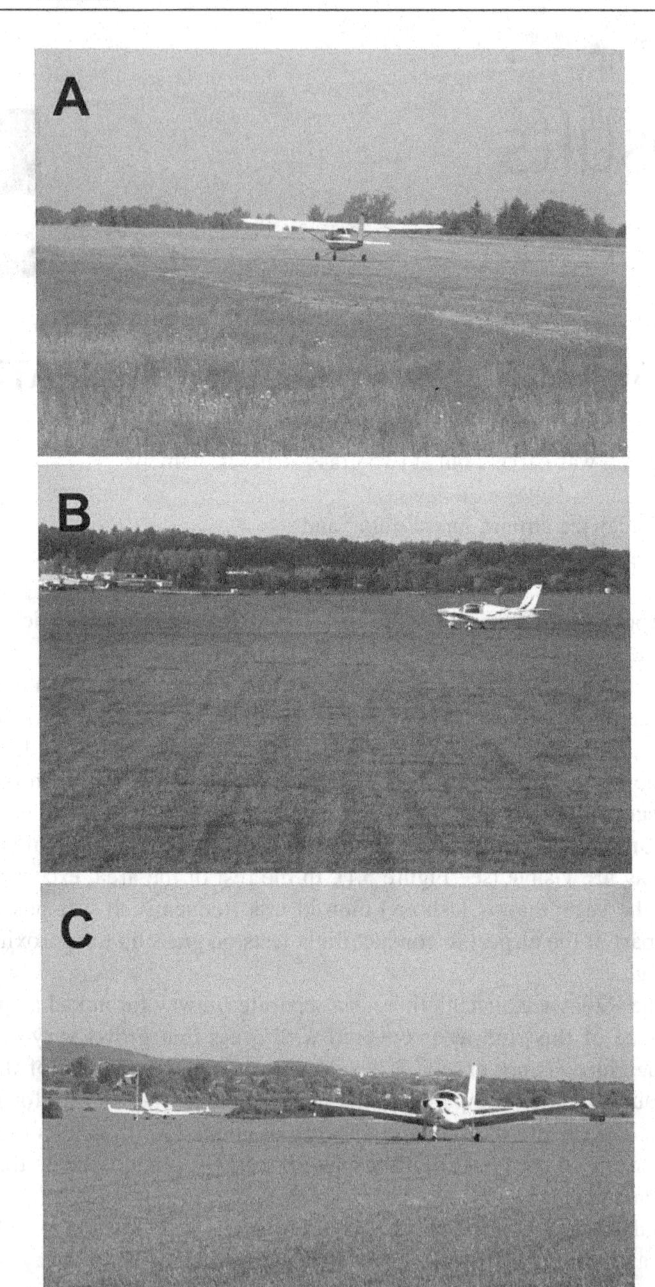

FIGURE 5.1 Grass airfields where the research was conducted. A – Radawiec, B – Świdnik, and C – Depułtycze Królewskie.

Occasionally, measurements were also carried out on the grass runway adjacent to the Lublin Airport, located in Świdnik, near Lublin. The soil substrate is very similar to that in Radawiec, but the grass growing on this runway is characterized by a more homogeneous composition, without the meadow vegetation. The grass on the runway of this airport is mowed and approximately is 5–7 cm high. It is rather rarely growing, partially degenerate. The rest of the airport is covered with a fairly dense mixture of grasses and meadow plants. At the Świdnik Airport, there is a colony of pearl gopher, and this species is completely protected. The negative effect of the presence of this animal is the formation of unevenness, depressions, and cavities in green vegetation.

In general, selected grass airfields are typical in terms of soil and plant conditions for the area of Central and Western Europe. Similarly, in terms of weather conditions, they are typical of the temperate zone, with little influence from the continental climate, especially in summer and winter.

Figure 5.1 shows the view of the grass runways where flight tests were carried out.

5.1.1 Cone Index

The so-called full-season (year-round) penetrometric measurement has been performed at the two grass airfields. The measurement was performed monthly, trying to select the time of measurement (day) with the weather typical for a given month in the year. During that time, the soil Moisture Content (MC) was measured. The results of penetrometric measurements for both grass airfields were correlated with the results of measurements of soil MC. The correlations obtained in this way are presented in Figure 5.2.

The CI values measured at a depth of 15 m are shown here. Relationships obtained are of exponential character for both airfields. However, it can be seen that the ranges of values are different, namely at the Depułtycze Królewskie airport, a wider range of CI values was recorded, reaching values below 1,000 kPa, while in the grass runway of the Radawiec airport, the minimum recorded values were approximately 2,500 kPa. The difference is due to the type of soil which is the ground of the grass surface and the fact that the turf of the grass runway in Depułtycze was built recently and is not subject to mechanical rolling. The soil is looser and less compacted. Additionally, the root system in the soil at Radawiec airport is strengthened thanks to the diversity of vegetation (grass + meadow vegetation).

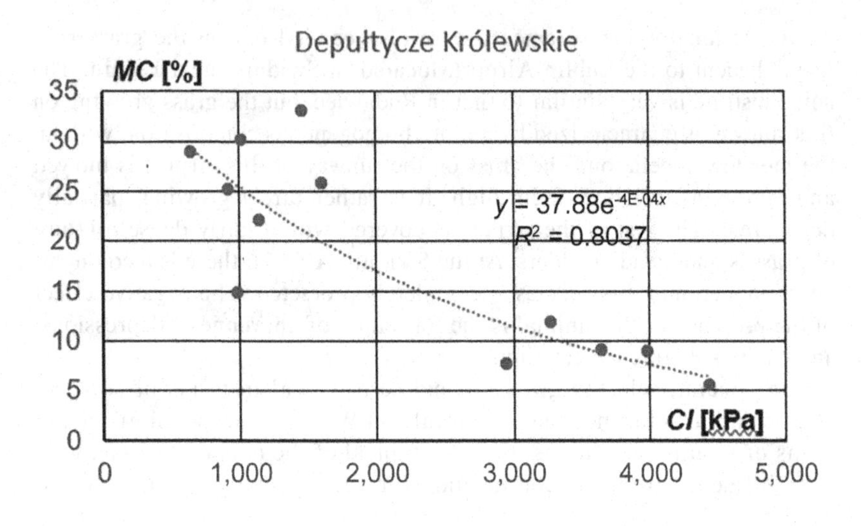

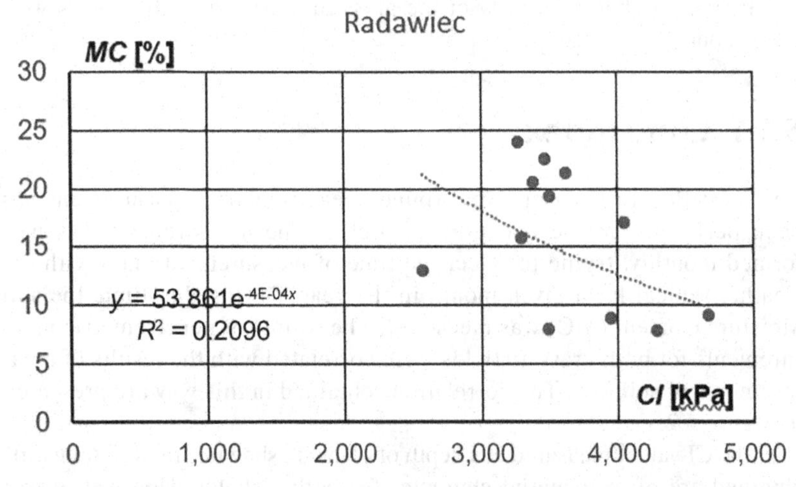

FIGURE 5.2 The relationship CI–MC (Cone Index–Moisture Content) obtained for soils of two grass runways: Radawiec and Depułtycze Królewskie.

Penetrometric measurements were also carried out during flight tests on the runway of a given airport, along its axis. Performed minimum 12 repetitions, the measurement sites were chosen randomly, going from the beginning of the runway, following the takeoff/landing direction. Simultaneously with the penetrometric measurement, the measurement of soil moisture was performed, and samples of green parts of vegetation were taken from an area of 1 m^2.

5.1.2 Rolling resistance of grass runway

Figure 5.3 shows the results of the rolling resistance force F_{RR} and the rolling resistance coefficient k_{RR} for the wheels of the PZL 104 Wilga 35A aircraft, obtained from dynamometric tow measurements carried out at the Radawiec Airport under various conditions. The lowest values of both rolling resistance forces and the k_{RR} coefficient were obtained on the model airplane runway.

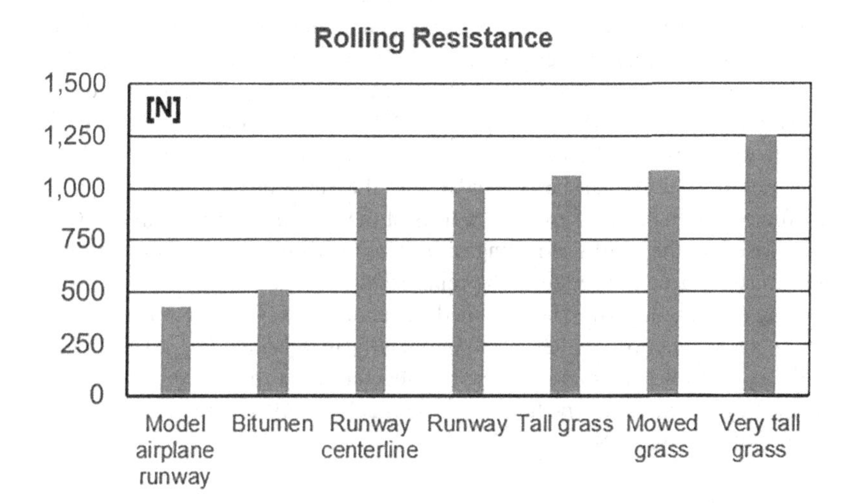

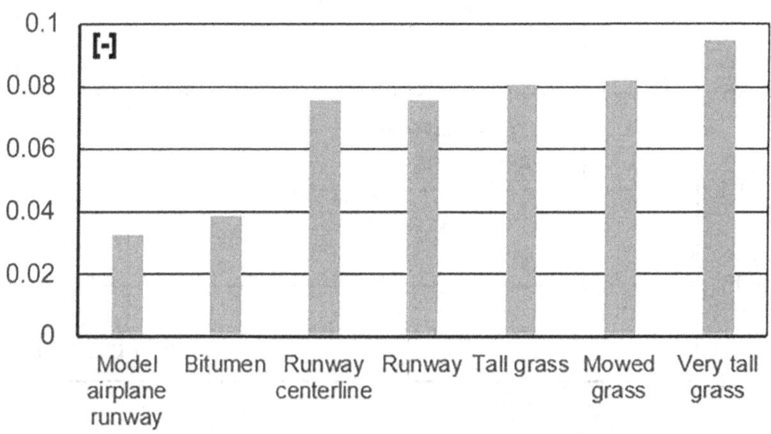

FIGURE 5.3 Rolling resistance of aircraft tires measured by means of pull test method.

On the concrete surface, the measured F_{RR} and k_{RR} values were slightly higher (apron in front of the hangar). This can be explained by the fact that the wheels of the PZL 104 Wilga 35A aircraft have low-pressure tires, specially adapted to the grass belt. Thus, low-pressure aircraft tires do not perform better on a concrete pavement in terms of rolling resistance. The low-pressure tire deflects significantly on a rigid concrete pavement, a large contact area is built up, and the resulting rolling resistance is high. The next results presented in Figure 5.3 concern the grass surface, whereby the surface of the runway was distinguished from the rest of the airfield surface. Additionally, the results for the surface with mowed, wet grass are included. It should be noted that the highest values of F_{RR} and k_{RR} were recorded for very tall grass (approximately 15–20 cm) [1, 2]. In accordance with the general recommendations and in accordance with the flight manual for the PZL 104 Wilga 35A aircraft, flying on such a surface of the runway is barely allowed.

Summing up, it can be concluded that the value of the rolling resistance coefficient for the wheels used in the tests of the aircraft fluctuates around 0.08.

Based on the results of rolling resistance measurements and the results of unit mass of green vegetation, the relationship k_{RR}–m_G was drawn up, which is shown in Figure 5.4. The obtained dependence was approximated by the exponential function, which is monotonically increasing. The result is logically explainable: the more intense the vegetation on the runway, the greater the rolling resistance of the landing gear wheel, as more energy is lost to compacting the green parts of grass.

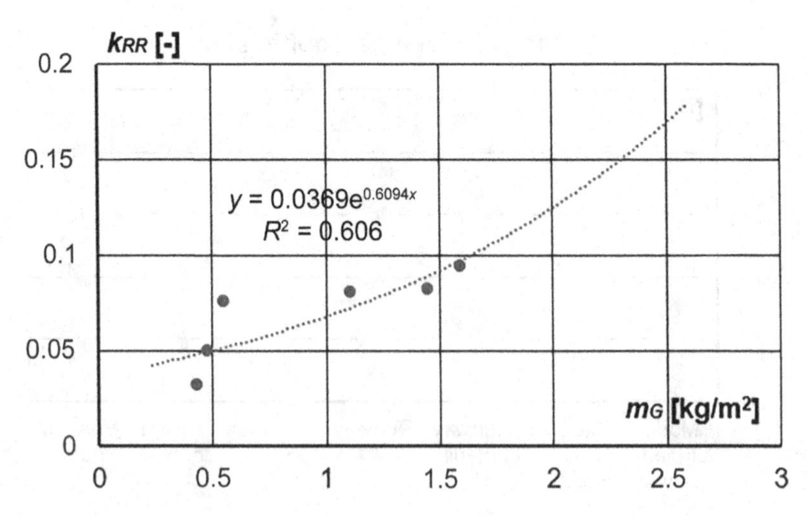

FIGURE 5.4 Rolling resistance coefficient related to unit mass of green vegetation.

5.1.3 Effect of soil moisture content upon rolling resistance and braking friction

Next results were obtained in measurements using an instrumented aircraft [3]. The forces and moments acting on the landing gear wheel of the PZL 104 Wilga 35A on which the dynamometer was installed were measured. Measurements were carried out on a grass runway with sparse vegetation at various values of soil MC. To determine k_{RR} and μ, the following values were used: the moment acting around the lateral axis M_Y, the wheel load force F_Z as well as the dynamic radius of the tire, r_d. The rolling resistance coefficient is determined from the relationship:

$$k_{RR} = \frac{M_Y \cdot r_d}{F_Z} \tag{5.1}$$

for a range where M_Y values are stable.

The M_Y^{max} value, obtained when the instrumented vehicle was stopped (wheels braked), is used for further calculation. Then, the braking friction coefficient is determined with the use of the following equation:

$$\mu = \frac{M_Y^{max} \cdot r_d}{F_Z} \tag{5.2}$$

Figure 5.5 contains a set of values of the k_{RR} and μ coefficients for various values of soil moisture in the grassy runway substrate. The rolling resistance coefficient increases with increasing humidity, and this relationship has been described by an exponential function, which is characterized by a very good fit to the results obtained in the measurements. Regarding the braking friction coefficient, the opposite is true: μ decreases with increasing soil moisture. Also in this case, the exponential function was used to mathematically describe the relationship, the fit being slightly worse than in the case of k_{RR}. In both cases, the obtained dependencies are logically justified. The increase in soil moisture causes a decrease in its mechanical strength, as a result of which the deformation of the soil surface is greater and the resulting rolling resistance increases. In the case of μ coefficient, an increasing soil moisture causes its plasticization and greater tendency to shear deformation. In the tire tread–soil contact area, shearing of the top soil layer takes place, and, moreover, the rubber–soil adhesion is weaker due to the water that wets the soil particles, which translates into lower surface tension and less adhesion.

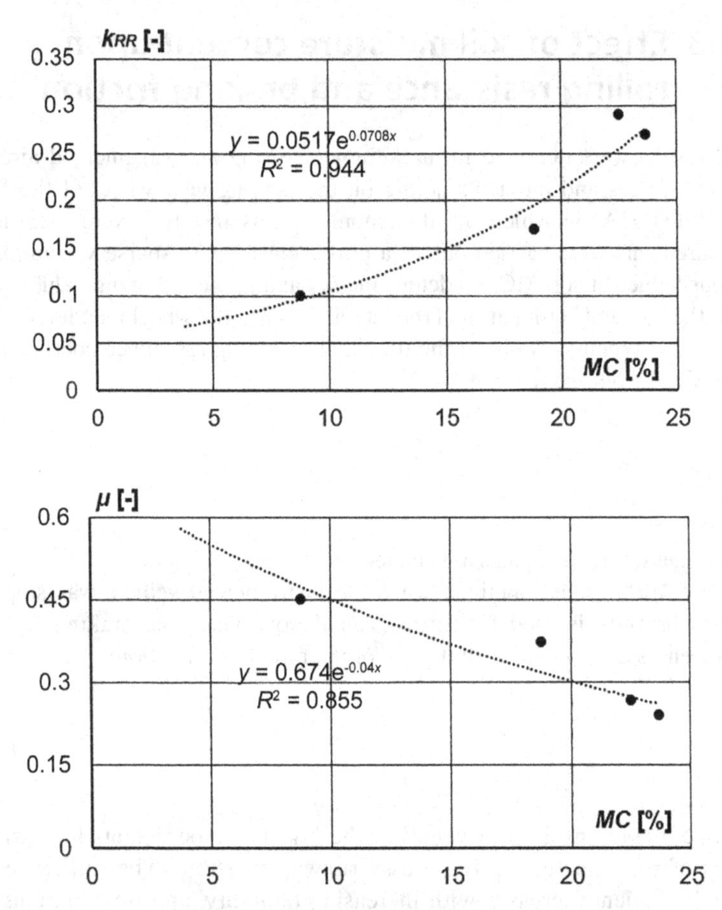

FIGURE 5.5 Braking friction coefficient as a function of soil moisture content for short grass.

5.2 EFFECT OF SPEED ON ROLLING RESISTANCE COEFFICIENT[1]

The effect of rolling speed on the k_{RR} has been determined using the method described in Section 4.3.1. The results of flights, that is, the increase of airplane ground speed as a function of time, were converted into accelerations, and the values of the k_{RR} coefficient for each speed range were determined. Final results are shown in the form of k_{RR}–V relationships in Figure 5.6 for

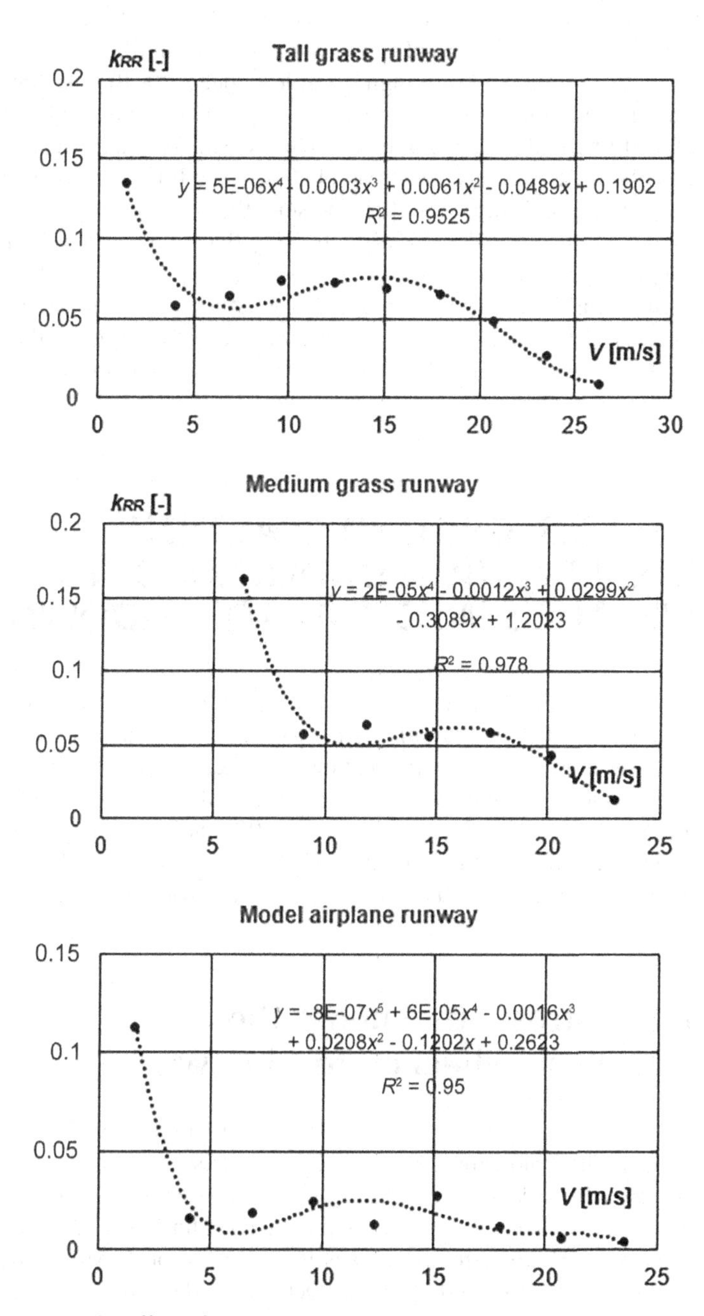

FIGURE 5.6 The effect of airplane ground speed upon rolling resistance coefficient.

the three investigated surfaces: tall grass, medium grass, and the model airplane airfield. A general shape of the k_{RR}–V curves is similar for all three cases: decreasing from a relative high k_{RR} at low speeds (0–10 m/s) down to a local minimum point at about 10 m/s, then increasing in the middle range of speed, and finally decreasing at higher speeds. This is in agreement with the results obtained by Crehshaw (1972) [4]. There is a little or no difference between the two grass surfaces, while the smoother model aircraft runway differs significantly from the others with a k_{RR} that is much lower in the whole range of ground speeds. Both grass surfaces were rough, while the model airfield was smooth (this is a subjective estimation by the author, as the roughness of the surfaces was not determined). This suggests that the total rolling resistance is also dependent on roughness, and not only on the length of the grass blades [3].

5.3 GROUND REACTION DETERMINATION UNDER LOADING OF AIRPLANE'S WHEEL AT TOUCHDOWN[2]

The Stress State Transducer (SST) soil stress sensor was placed at a depth of approximately 10 cm in the soil of grass runway of the Radawiec airfield. The place was marked with a white rectangle marker. The pilots who participated in the research, got trained in the landing skills in preparation for the precision flying championships. More than a 100 landings were performed and 32 were successful – the left wheel of the aircraft touched the ground at the point above the SST, but only some of them provided satisfactory soil pressure data.

5.3.1 Soil stress state under the airplane wheel on touchdown

Figure 5.7 shows the courses of stresses in the soil while Table 5.1 contains numeric data. The major stress, σ_1, was the dominant stress component for all tests. For zero-thrust landings, a quasi-two-dimensional stress state ($\sigma_2 = 0$) was observed there, while for normal landings and landings, where the aircraft rolled over the SST, the stress state was three dimensional. Minor stresses reached 15–30% of maximum σ_1 values. Octahedral shear stress was always greater than the mean normal stress.

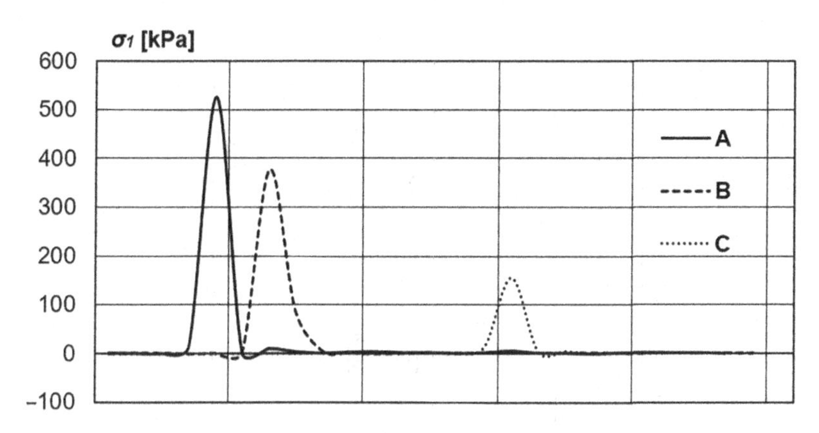

FIGURE 5.7 Soil reaction to the load of airplane wheel at touchdown.

The measured values of the stress state components were significantly higher for normal than for zero-thrust landings. The difference was more than 20% for σ_1, so the effect can be regarded as significant. Moreover, the time in which the stresses are active was longer for zero-thrust landings (0.03 s versus 0.02 s). The zero-thrust landings are typically softer, as the pilot has no possibility of engine control, and the forces acting on the aircraft are stable (only when the landing is technically right, however). In cases of normal landings, the propeller thrust force is a significant component that influences ground reaction and, furthermore, the soil stress state components. On final approach, the power setting was about 20–30%, and, for approximately 1 s before touchdown (flare), the engine was set on idle power, which causes a sudden sinkage and the airplane hits the surface with a relatively high vertical speed, since the pilot wanted to touch the ground in the marked point. On the other hand, the constant idle-power setting by zero-thrust landings required the pilots to utilize the greatest glide ratio of the aircraft.

In a normal landing, the airplane is more likely to hit the surface, but in a zero-thrust landing, the flight profile is gentle, and the wheels touch the surface softly. Another factor that affected the soil stress state in this experiment was the technique used during the landings.

Rolling over the SST with a high longitudinal velocity (or landing velocity) generates a soil stress state in which components are significantly smaller than for a touchdown. An impact effect and the gravity force are partially balanced by aerodynamic lift force, so the contact pressure and soil stress are smaller.

TABLE 5.1 Soil stress state data for conventional landings, emergency zero-thrust landings, and rolling over the SST[3]

		STRESS STATE COMPONENTS (KPA)				
LANDING PATTERN	*NO. OF TEST*	*S1*	*S2*	*S3*	*MNS*	*OCTSS*
	48	416.9	283.0*	−347.0*	117.6*	333.1*
Zero-thrust	110	375.9	0.3958	−248.2	42.71	256.5
landings	207	382.0	−0.7585	−253.4	42.63	261.2
	Average/std dev	391.6/18.06	−0.577/0.18	−250.08/2.6	42.67/0.04	258.85/2.35
	121	526.1	94.62	−219.7	133.7	305.7
Conventional	203	506.4	29.66	−4.847	177.1	233.3
landings	Average/std dev	516.25/9.85	62.14/32.48	–	155.4/21.7	269.5/36.2
	111	155.5	−0.2112	−42.27	37.67	85.06
Rolling	119	165.4	−13.10	−50.09	34.05	94.06
over the SST	202	121.0	0.5102	−40.86	26.87	68.64
	Average/std dev	147.3/19.03	−4.26/6.25	−44.38/4.03	32.86/4.48	82.58/10.52

Note: Values marked with "*" were ignored as being over-registered

5.3.2 Orientation of the inertial force and σ_1

Having captured the acceleration data of the entire aircraft in the moment of the touchdown, it was possible to determine both inertial force and orientation. Vertical, longitudinal, and horizontal (i.e., transversal) acceleration data were collected. The values are the average of the five repetitions. Both vertical and horizontal accelerations are significantly greater for normal landings, and this trend corresponds to the σ_1 peak values. The orientation of the acceleration-inertial force vector, however, was not affected by landing patterns.

The SST- measuring method allows the determination of both soil stress state components, as well as their direction cosines. The orientation of the stress state components, especially the major stress, σ_1, is a valuable information, as the character of the loading suggests that the σ_1 is not vertical at the location where it was maximum. In such a case, minor stress components are of high importance, and the stress state is three or two dimensional. These data can be related to the orientation of the inertial force, which is the major force generating stresses in soil.

The results for δ (the tilt angle of σ_1) and γ (the tilt angle of inertial force) were obtained for two landing patterns and for the case where the aircraft rolled over the SST (see Figure 5.8). The landing pattern has a significant effect on the σ_1 orientation: for a maximum value of the σ_1, the stress vector is pronouncedly more tilted for normal landings. The σ_1 tilt angle is much smaller for the rolling over the SST case. In any case, the relatively high values of δ for both landing patterns suggest that the other stress state components should be analyzed to be significant. Peak minor stresses σ_2 and σ_3 range from almost 0 (σ_2 for zero-thrust landings) to –250 kPa (σ_3 for the same) and 62 kPa (σ_2 for normal landings). However, we observed no significant trend in σ_2 and σ_3 peaks for different landing patterns.

Apparently, σ_1 is more tilted than F_Q for the various landing patterns. This may be caused by the kinematics of the undercarriage: the main wheels are castored and supported by an oleo shock absorber, so that vertical movements of the wheels are exaggerated.

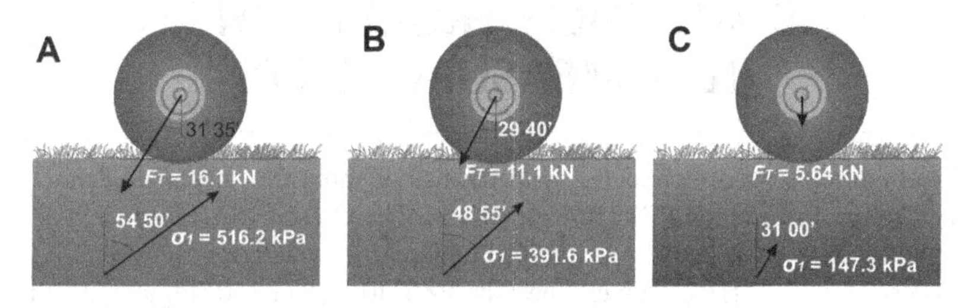

FIGURE 5.8 Orientation of soil reaction force.

5.3.3 Concluding remarks

Would the approach presented earlier be practically useful? If so, what are the limitations of the method? First, the assumption of the elasticity theory needs validation. Stress state data obtained for soft, loose soil; fresh snow; or till should be analyzed with care. It is insufficient to relate the stress values to deformations without the analysis of nonlinearity. Konstankiewicz investigated stress–strain relationships for sandy and loamy soil for various deformation velocities; a nonlinear parametric model gave promising results when compared with the experiments. Other promising methods are stochastic or probabilistic models [12].

Second, the repeatability of the experimental results and the methodology of stress measurements are at issue. The probability of a successful landing trial, with a touchdown taking place exactly over the SST, was below 10% in the present experiment. Therefore, more than a 100 trials were needed to obtain results that are of scientific value. Moreover, the reliability of the obtained results is below 90%, as we were not completely sure where the wheel hit the surface (a photo or video technique would be helpful). It would be reasonable to use a 2×2 or 3×3 matrix of SSTs in a future test. Another factor is the landing technique. To obtain the highest probability of success, champion pilots were engaged for the experiment. The pilots used a special competition technique in which the precision of the landing is most important. In such landings, abrupt maneuvers on approach and intensive thrust control lead to semi-forced landings that were not symmetrical. Such landing technique is not a handbook standard, and it would be interesting to see how different the standard landings would be.

Concluding, the soil stress–strain-based method seems to be convenient, but more research should be conducted in the field of measuring methods to improve accuracy and repeatability.

5.4 RESULTS FROM FLIGHT TEST MEASUREMENTS

The flight tests were conducted with the use of three GA airplanes, namely the Cessna 152 and the PZL110 Koliber 150. The Cessna 152 (Figure 5.9A) is a high-wing, two-seater, three-wheeled undercarriage with a nose wheel. It is powered by a 120 HP piston engine. The PZL110 Kolber 150 (Figure 5.9B) is a low-wing airplane with a four-seat cabin. The plane is powered by a 150 HP

FIGURE 5.9 Airplanes that were used for flight tests: Cessna 152 (A, upper photograph), PZL110 Koliber 150 (B, middle photograph), and PZL 104 Wilga 35A (C, lower photograph).

engine. The PZL 104 Wilga 35A (Figure 5.9C) is a high-wing four-seat airplane with a taildragger-type landing gear. The airplane is powered by a radial piston engine of 260 HP power.

The main purpose of the flight tests was to determine the dependence of the airfield performance – the length of takeoff and landing ground roll – on important factors such as vegetation, soil strength, soil moisture, wind, and density altitude. The IMUMETER device was used for the measurements [5–7]. The flight test campaign lasted from June to October 2022. Approximately 10–12 repetitions were performed for each test. The results were averaged, and highest/lowest values were rejected. The obtained results were used to prepare the correlation and to identify the models describing the airfield performance of the aircraft on a grass runway.

5.4.1 Effect of soil strength

Figure 5.10 shows the relationship between takeoff and landing ground roll for the PZL110 Koliber 150 aircraft and the CI values. The soil, which is the base of the runway on which the measurements were made, is a compacted

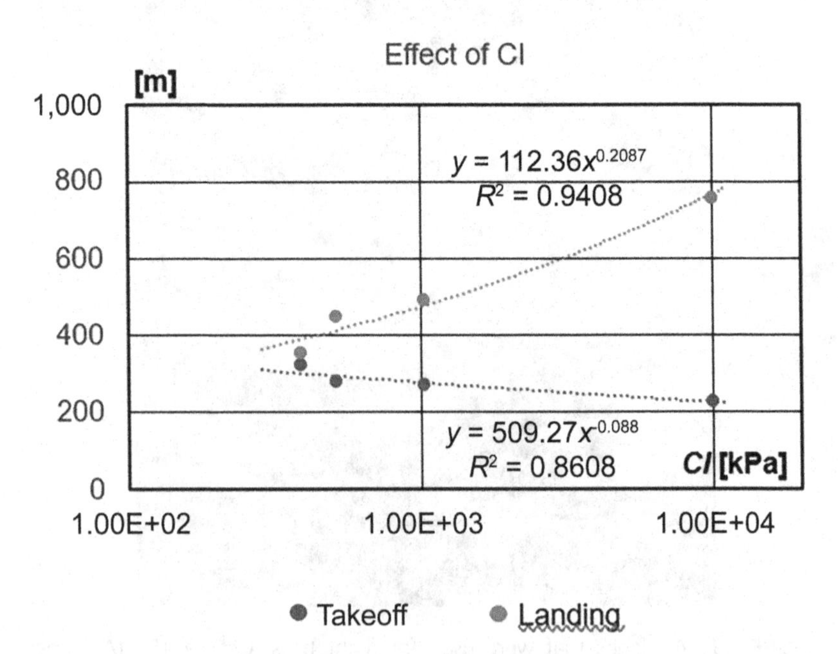

FIGURE 5.10 The effect of Cone Index on takeoff and landing ground roll of the PZL110 Koliber 150 aircraft.

loess-sandy soil. The CI effect is different for takeoff and landing distances. The length of the takeoff roll decreases with increasing CI, while the length of the landing roll increases. It is related to the reduction of the rolling resistance as the CI increases. Since the wheel brakes were not used during landing, the free rolling of the airplane was longer when the plane landed in high-strength soil conditions. The mathematical models describing the studied effect are of the exponential type and are characterized by good ($R^2 = 0.86$ for takeoff) or very good ($R^2 = 0.94$ for landing) fit to the experimental data. The tests were carried out in sparse vegetation conditions to minimize the influence of grass.

Thus, it is concluded that the soil strength measured with the cone penetrometer is a factor that influences the ground performance during takeoff and landing of an airplane on a grass runway. Further research should focus on the analysis of the influence of this factor on the different types of soil that underlie the grass runway.

5.4.2 Effect of soil moisture

Another factor related to the characteristics of a grass runway is soil moisture. Figure 5.11 shows the dependence of the takeoff and landing ground roll of the PZL110 Koliber 150 aircraft upon soil moisture. As before, the measurements were made under sparse vegetation conditions. There is a clear progressive effect of soil moisture on the length of takeoff and a degressive effect on the

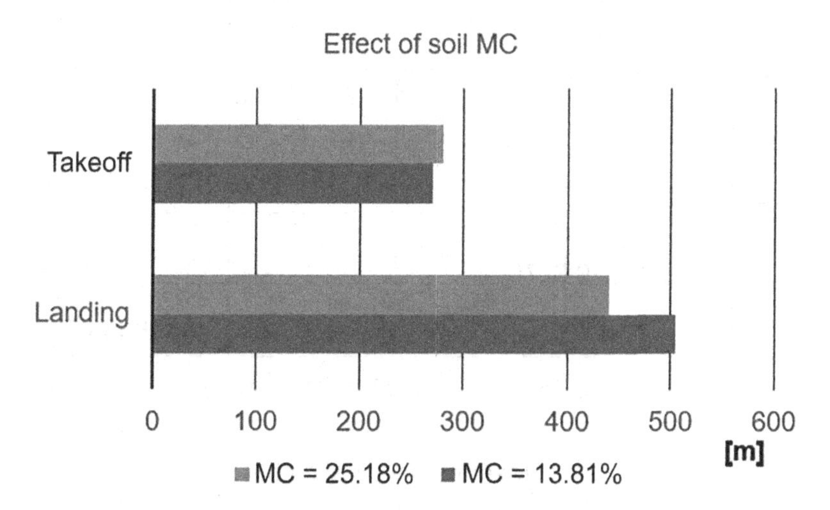

FIGURE 5.11 The effect of soil moisture content on takeoff and landing ground roll of the PZL110 Koliber 150 aircraft.

length of landing. However, the obtained dependencies differ in terms of intensity: the reduction of the landing distance is much more intense (14%) than the increase in the takeoff distance (3.5%). The obtained results are logical, higher soil moisture results in a decrease in its strength and, consequently, in greater soil deflection under the load of the landing gear wheels. The final effect is a higher rolling resistance, which translates into a longer takeoff and a shorter landing.

5.4.3 Effect of vegetation

The influence of green parts of grass on the airfield performance of the PZL110 Koliber 150 aircraft is shown in Figures 5.12 and 5.13. The dependence of takeoff and landing distances upon the m_G parameter for dry conditions (low soil moisture, high air temperature) is shown in Figure 5.12. In the case of takeoff, the ground distance increases, which is due to the higher rolling resistance when the vegetation is more intense. Similarly, when the landing rollout is decreasing, the airplane is additionally braked by more green parts of the grass. The obtained dependencies are of the exponential type, with a very good coefficient of matching to the measurement results. Figure 5.13 shows

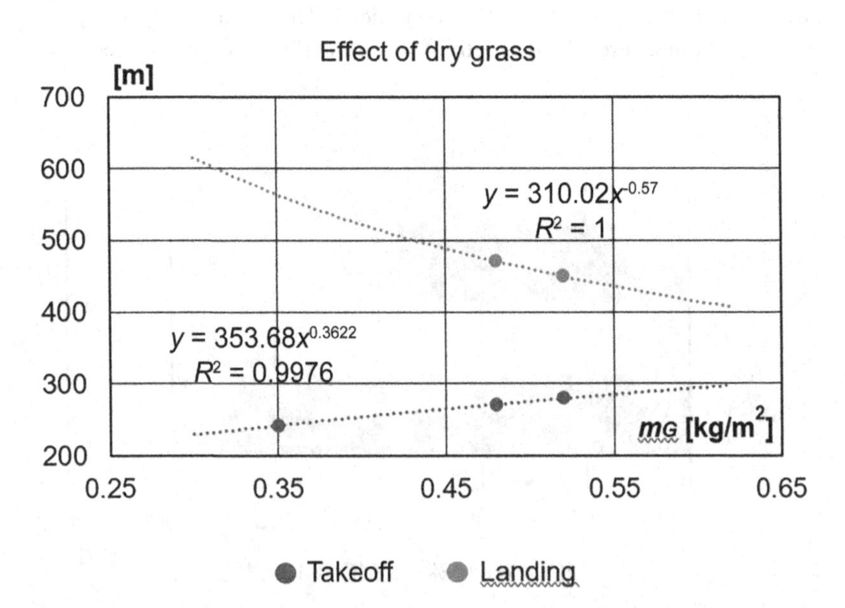

FIGURE 5.12 The effect of vegetation on takeoff and landing ground roll of the PZL110 Koliber 150 aircraft at dry soil conditions.

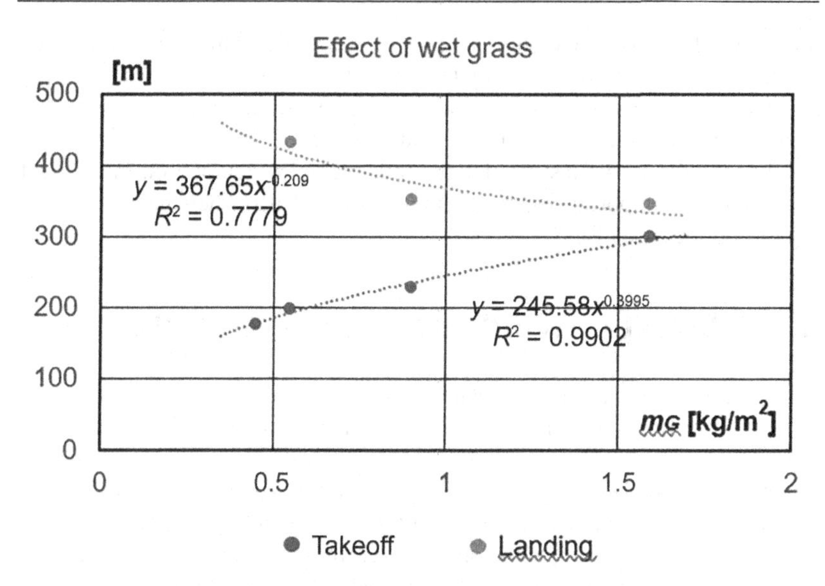

FIGURE 5.13 The effect of vegetation on takeoff and landing ground roll of the PZL110 Koliber 150 aircraft at wet soil conditions.

the relationship among vegetation and takeoff and landing performance of the Koliber airplane for wet soil conditions (soil moisture in the range of 23–30%). The mathematical model describing the effect of vegetation on takeoff ground roll distance is of the exponential type, characterized by a high value of R^2, which is 0.99, while the model describing the relationship between the m_G and landing ground roll is also of the exponential type, characterized by a satisfactory fit to the measured values ($R^2 = 0.77$). During these tests, the brakes on the landing gear wheels were not used. From the results, it can be seen that the takeoff is longer and the landing is shorter as the m_G increases, as in the case of dry grass, but the wet grass effect is much more pronounced. Both the increase in the takeoff distance and the shortening of the landing ground roll are more intense on the runway with wet grass. While in the case of a takeoff, such result is logical and expected, in the case of a rollout, one would expect that the reduced traction on a wet grass runway would result in a lengthening of the landing run. This would be the case if the brakes were used during the tests.

The influence of vegetation is clear. The obtained relationships are logical and explainable: for increasing m_G, the rolling resistance increases, which translates into an increased takeoff ground roll. Similar results were obtained by Shoop et al. (2015) as well as Wieder and Shoop (2018) [8, 9]. On the other hand, the decrease in the rollout is caused by an increase in rolling resistance, owing to which the plane stops at a shorter distance. This increase of rolling resistance is mainly caused by the conversion of the kinetic energy of the airplane rolling on the runway surface into the work necessary to knead and compact the grass under the landing gear wheels.

In the light of the test results, the influence of vegetation on the airfield performance is important especially for takeoff and also when the grass covering the runway is moist.

5.4.4 Effect of wind and density altitude

Two weather factors, wind and density altitude, are important from the point of view of airfield performance during takeoff and landing. The influence of the wind was determined for the Cessna 152 aircraft. It is the lightest aircraft among those used for the research. Figure 5.14 shows the result, that is, the relationship among takeoff and landing ground roll and wind speed. It should be noted that the average wind speed given in the weather report was used for the analysis (the wind was not measured). If the wind direction differed from the takeoff and landing direction, a component parallel to the direction of flight of the airplane was determined. Analyzing the obtained results, it is stated that the wind has a significant importance in terms of the takeoff

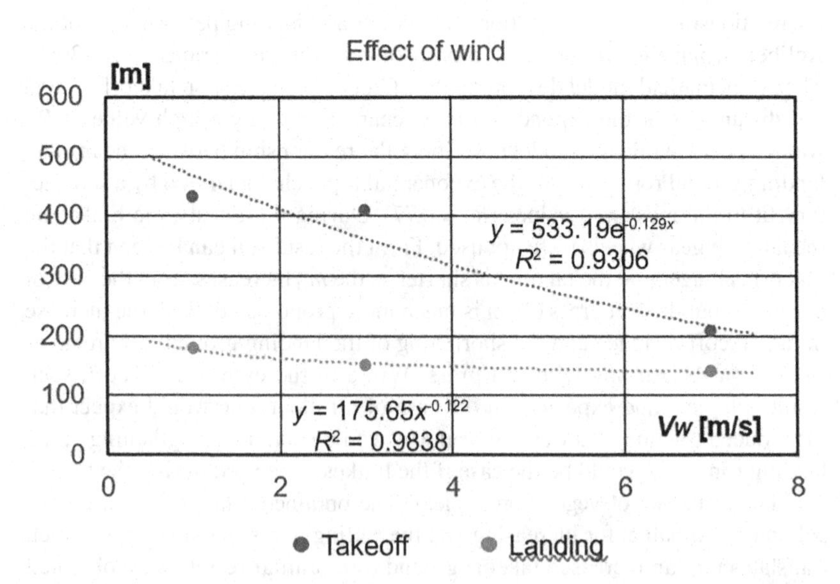

FIGURE 5.14 The effect of wind on takeoff and landing ground roll of the Cessna 152 aircraft.

ground roll: the higher the wind speed, the shorter the takeoff. The mathematical model reconstructed from the experimental data is of the exponential type and is characterized by a satisfactory fit to the measurement results. This result is logical and expected. Similarly, in the case of landing, it was found that the wind speed had a degresive effect on the landing ground roll, although this effect was less intensive.

The second weather factor, that is, Density Altitude (DA), indirectly influences airfield performance in two ways. First, the change in air density resulting from changes in atmospheric pressure and outside air temperature implies a change in engine power, which translates into airfield performance, especially at takeoff. Second, air density directly affects the aerodynamic forces, including the lift and drag forces. As the air density changes, so does the liftoff speed and therefore the takeoff ground roll. The expected impact of density altitude on the rollout is negligible. Figure 5.15 shows the dependence of the takeoff run on the density altitude obtained for the PZL110 Koliber 150 aircraft. It turns out that the influence of air density is very significant, as the increase in density altitude (DA) by 240 ft caused the PZL110 Koliber's takeoff to lengthen by about 40 m. The mathematical model describing this relationship is an exponential function which, in relation to the measurement results, is characterized by a regression coefficient R^2, which is equal to 0.95.

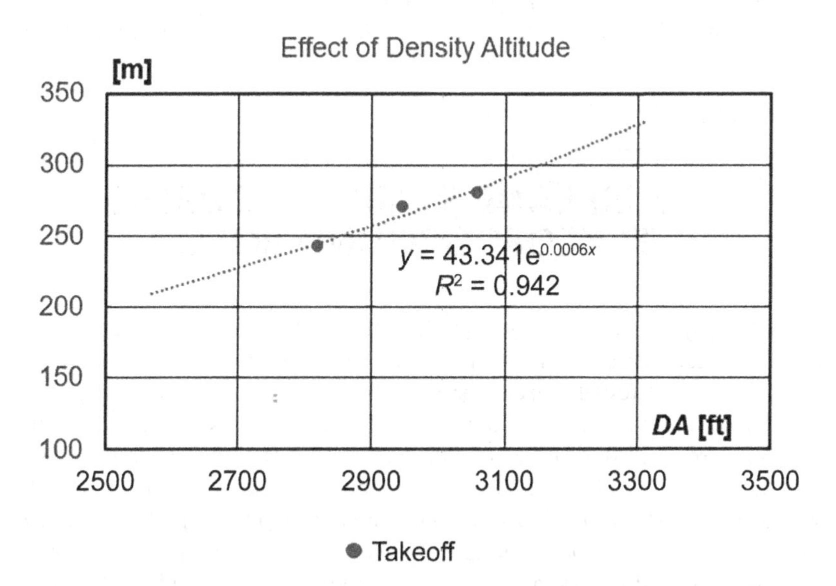

FIGURE 5.15 The effect of density altitude on takeoff and landing ground roll of the PZL110 Koliber 150 aircraft.

5.4.5 Conclusions

The relationships among the takeoff and landing ground roll and the parameters characterizing the grass runway – soil CI, soil MC, and the mass of green vegetation per unit of runway area presented in this chapter – are the main achievements of the whole study. The obtained dependencies do not cover the full range of values, taking into account the parameters; additionally, they are in many cases obtained on the basis of a small amount of data. This is due to the simultaneous influence of several factors and the high dynamics of the conditions in which the measurements were carried out. It is impossible to control weather factors. Conducting measurements in repeatable conditions was difficult, mainly due to the influence of weather factors (wind, air density). Waiting for a change in soil moisture resulted in a simultaneous change in altitude density. Often times, the wind was variable, although attempts were made to take measurements in the morning. All this meant that not all the obtained measurement results could be used for the construction of the relationships describing airfield performance. The conclusion that arises is that it would be advantageous to isolate a fragment of a large grass airfield, for example, Radawiec, and maintain an experimental runway with controlled vegetation, owing to which it will be possible to determine one of the important research parameters. It is also necessary to efficiently cooperate with the institution lending the plane with the crew. Thanks to the help of the Świdnik Aeroclub, it was ensured at the highest level.

5.5 AIRPLANE STABILITY DURING TAKEOFF GROUND ROLL

The DGPS system, also known from the research on the ride dynamics of off-road vehicles, measures a number of kinematic quantities characterizing the movement of the tested object. These are linear velocities and accelerations, angular velocities and accelerations, position in space, sideslip angle, and others. As there is no standardized method for testing the stability of the plane motion on the ground, in the form of a standard or other formal document, it was decided that the stability test would be carried out on the basis of three quantities used in vehicle dynamics research [10]. These are the lateral acceleration, the side slip angle of the center of gravity, and the angular velocity of the yaw movement.

The reduction of the experimental data consisted in choosing the series containing the values of the lateral acceleration, the side slip angle, and the yaw rate for selected ranges of the aircraft speed: 10, 20, 30, 40, 50, 60, 70, 80 and 88 km/h. The last speed value basically relates to the liftoff point (it varies, 88 km/h is the average value). For each range of aircraft speed, 50 measurement points were distinguished so that the beginning of the series began at a speed of about (V_i – 1.5 km/h) and ended at (V_i + 1.5 km/h), for example: (48.5–51.5 km/h). The series have been prepared in this way for both cases (short and tall grass) and were used for the identification procedure with the use of the so-called small disturbances method.

5.5.1 Effect of short grass

Figure 5.16 shows the results obtained on short grass and includes root locus plots for the three dynamics parameters: the lateral acceleration, the sideslip angle, and the yaw rate. Unstable roots are marked in red. When analyzing the locus roots for individual speed ranges, it is visible that for lower speeds (0, 10, 20, 30, and 40 km/h), the aircraft motion described in term of lateral acceleration is stable.

At speeds in the range of ca. 50 km/h, the motion described in term of lateral acceleration becomes unstable, while there is a fairly clear upward trend in the measure of instability with increasing speed, and the greatest instability occurs at the liftoff speed (approximately 88 km/h). It is interesting, however, that at a speed of 70 km/h, the motion of the plane is stable. A logical explanation of this phenomenon (the return of stability at a certain speed) may be the effect of lifting the tail wheel. During the takeoff run, aerodynamic forces begin to act, including lift, which causes levelling the plane. From then on, the aircraft touches the ground only with the main landing gear wheels. Perhaps this change in the landing gear configuration caused a temporary return to the state of stability. On the basis of computer simulation results, Abzug (1999) analyzed the moment when the aerodynamic forces responsible for the dynamics and lateral stability exceed the forces in the wheel–ground system. For example, for the Cessna 182, the speed at which this happened was 40 km/h [11].

Figure 5.16B includes the results obtained when the aircraft motion was described in term of the sideslip angle. These results are similar to the case discussed before (lateral acceleration). A temporary return to stability occurred at 70 and 80 km/h. Similar trends were noted as a result of the analysis carried out for the yawing motion. To sum up, it can be concluded that on the surface covered with grass, low stability of motion is maintained at speeds of up to approximately 40 km/h; above this range, the motion of the aircraft is unstable, except for the moment when, due to the effect of increasing

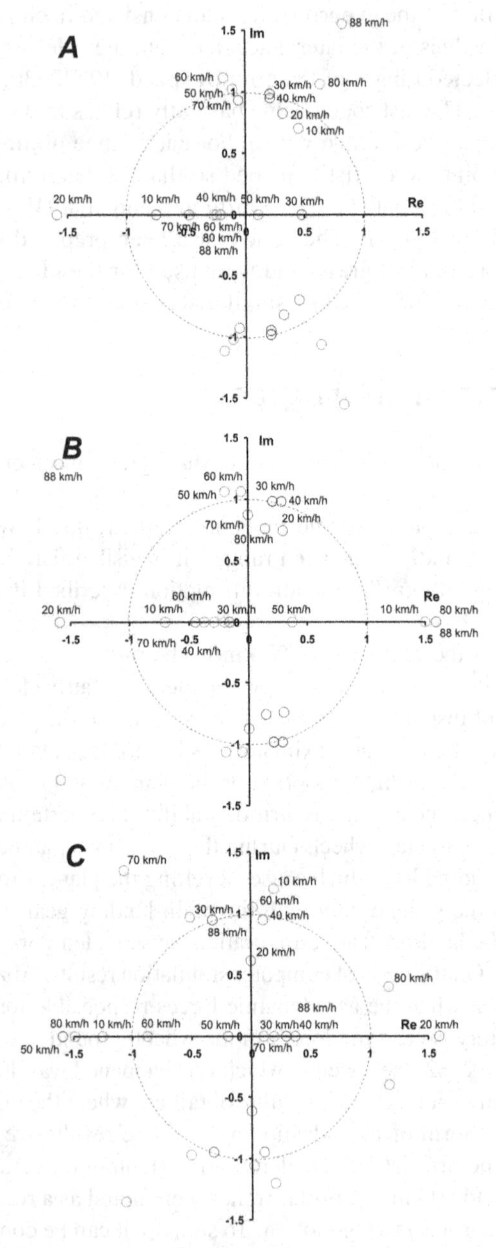

FIGURE 5.16 Locus roots of the characteristic equation of the model of the motion trajectory; A – lateral acceleration, B – side slip angle, C – yaw rate. The results were obtained on the runway covered with short grass.

aerodynamic forces, at a speed of approximately 70–80 km/h, the tail lifts off and the motion is continued on the wheels of the main landing gear.

5.5.2 Effect of tall grass

Figure 5.17 shows the results of the analysis of the locus roots for the second case – runway surface covered with tall (long) grass. When analyzing the results, it was found that in this case, the speed of the plane while moving on the ground increased to an average value of 97 km/h (on short grass it was about 88 km/h). It was related to the headwind effect, which on the test day on the surface covered with low grass was blowing at a speed of approximately 15 km/h.

Analyzing the positions of the roots for the lateral acceleration trajectory shown in Figure 5.17A, it can be concluded that the aircraft motion is unstable in this case. Momentary stability occurs at around 80 km/h. Similarly, in the case of the side slip angle (Figure 5.17B), the instability occurred only at 50 km/h and at high speeds (80 and 97 km/h – at liftoff speed). Stability occurs at 40 km/h. It follows that the stability of motion on tall grass is determined by the moment when the aerodynamic forces exceed the forces on the wheels and higher speeds (close to the liftoff velocity), but, generally, the aircraft motion is unstable.

5.5.3 Conclusion

The final conclusion is that the influence of the height of the grass is significant and the stability of the aircraft motion during rollout on short grass occurs for speed of 40–50 km/h, that is, the speed at which aerodynamic forces have a greater influence on the lateral dynamics than the forces in the wheel–ground system. On the other hand, on high grass, a significant instability occurs in the entire speed range, except in the case of stability at approximately 40–50 km/h, which may be the result of the aforementioned change in the domination of forces. Therefore, summing up, it can also be said that the effect of the change in the domination of forces is opposite for both cases, short and high grass.

5.6 SUMMARY

This chapter presented and discussed the results of experimental tests carried out with the use of research methods described in Chapter 4. The research

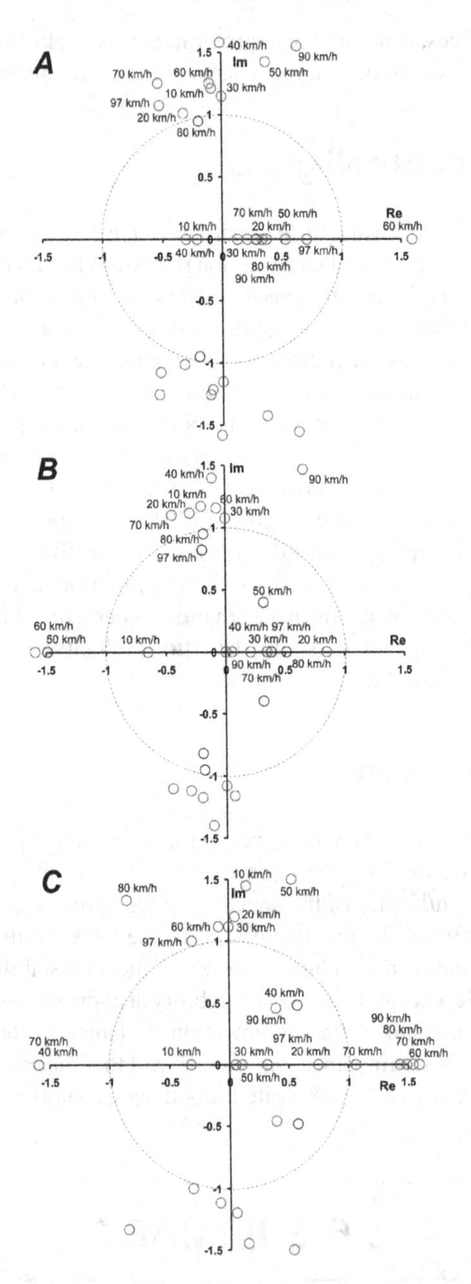

FIGURE 5.17 Locus roots of the characteristic equation of the model of the motion trajectory; A – lateral acceleration, B – side slip angle, and C – yaw rate. The results were obtained on the runway covered with tall grass.

used real objects – GA (General Aviation) airplanes –which, depending on the method used, were subjected to ground or flight testing. The presentation and discussion of the test results with the use of airplanes were preceded by the characterization of grass runway surfaces on which ground and flight tests were conducted.

The research focused on obtaining a description of the dependence of the airplane airfield performance upon the parameters characterizing the surface of the grass runway. The relationships of takeoff and landing ground roll distance with soil CI, soil MC, mass of green vegetation per unit of runway area, wind, and density altitude were obtained. It was found that all the aforementioned parameters of the grass runway significantly affect the takeoff and landing performance of airplane.

Moreover, the results of measuring the soil reaction in the runway soil substrate to the dynamic load from the landing gear wheel of the landing airplane were presented and discussed. It was found that the piloting technique influenced the magnitude of the soil reaction, measured by the value of the stress state in the soil. From the point of view of safety, the size of the horizontal stress component in the soil was considered to be significant.

The results of experimental and analytical studies of the stability of the aircraft movement during takeoff from a grass airfield are also presented. The speed ranges of the plane on the ground were determined in which there is a stable motion, and the speed of the plane at which instability occurs was also indicated.

One of the main goals of this work was to develop new methods for measuring the airfield performance of an airplane on a grass runway. The goal was achieved, and the methods developed were successfully applied in the research presented in this chapter. However, it is worth asking what to do next. According to the author, research in this field should be continued, especially to improve the safety of air operations and a more conscious use of grass airfields. It will be important to determine the effect of soil type and grass species on aircraft performance. Based on this, it will be possible to formulate practical recommendations for builders and users of airfields with grass runways. The answers to the questions of what type of soil and which grass mixture is the best from the point of view of lower rolling resistance and higher braking friction as well as the lowest possible sensitivity to the influence of weather conditions may be the target of future research in this area.

Another benefit resulting from the continuation of the ground performance tests of the aircraft on the grass runway is obtaining recommendations for pilots in the field of safe and effective flights, also in the autumn and winter period. A proposal of a practical solution in this regard, a design of an information system on the state of the surface of grass airfields, is presented in the next chapter of the work.

NOTES

1 Reprinted by permission of the American Institute of Aeronautics and Astronautics, Inc., from [3]. Copyright © AIAA.
2 This chapter was published in *Journal of Terramechanics*, Vol 40, Pytka J., Tarkowski, P., Dąbrowski, J., Bartler, S., Kalinowski, M., Konstankiewicz, K. *Soil stress and deformation determination under a landing airplane on an unsurfaced airfield*, pp. 255–269. Copyright Elsevier and International Society for Terrain – Vehicle Systems, 2004 [1].
3 This table was published in *Journal of Terramechanics*, Vol 40, Pytka J., Tarkowski, P., Dąbrowski, J., Bartler, S., Kalinowski, M., Konstankiewicz, K. *Soil stress and deformation determination under a landing airplane on an unsurfaced airfield*, pp. 255–269. Copyright Elsevier and International Society for Terrain – Vehicle Systems, 2004 [1].

REFERENCES

1. Pytka J., Tarkowski P., Dąbrowski J., Bartler S., Kalinowski M., Konstankiewicz K. Soil stress and deformation determination under a landing airplane on an unsurfaced airfield. *Journal of Terramechanics*, 40, 2004, pp. 255–269
2. Pytka J., Tarkowski P., Dąbrowski J., Zając M., Konstankiewicz K., Karczewski L. Determining the ground roll distance of an aircraft on unsurfaced airfield. *Proceedings of the 10th European Conference of the ISTVS*, Budapest, Hungary, October 2006
3. Pytka J.A. Identification of rolling resistance coefficients for aircraft tires on unsurfaced airfields. *Journal of Aircraft, AIAA Journal of Aircraft*, 51(2), 2014, pp. 353–360
4. Crenshaw B. Soil –wheel interaction at high speed. *Journal of Terramechanics*, 8(3), 1972, pp. 71–88
5. Pytka J., Budzyński P., Józwik J., Michałowska J., Tofil A., Łyszczyk T., Błażejczak D. Application of GNSS/INS and an optical sensor for determining airplane takeoff and landing performance on a grassy airfield. *Sensors*, 19, 2019, p. 5492. doi:10.3390/s19245492
6. Pytka J., Budzyński P., Tomiło P., Michałowska J., Błążejczak D., Gnapowski E., Pytka J.D., Gierczak K. Measurement of aircraft ground roll distance during takeoff and landing on a grass runway. *Measurement*, 195, May 2022, p. 111130
7. Tomiło P. Metoda pomiaru długości startu i lądowania samolotu z wykorzystaniem sztucznych sieci neuronowych (A method of measuring the take-off and landing distance of an airplane with the use of artificial neural networks). PhD thesis, in preparation, Lublin University of Technology
8. Shoop S.A., Coutermarsh B., Cary T., Howard H. Quantifying vegetation biomass impacts on vehicle mobility. *Journal of Terramechanics*, 61, 2015, pp. 63–76

9. Wieder W.L., Shoop S.A. State of the knowledge of vegetation impact on soil strength and trafficability. *Journal of Terramechanics*, 78, 2018, pp. 1–14
10. Pytka J., Tarkowski P., Kupicz W. A research of vehicle stability on deformable surfaces. *Eksploatacja i Niezawodnosc – Maintenance and Reliability*, 15(3), 2013, pp. 289–294
11. Abzug M.J. Directional stability and control during landing rollout. *Journal of Aircraft*, 36(3), May–June 1999
12. Konstankiewicz K. The experimental verification of the mechanical model of soil medium. *Proceedings of the III European ISTVS Conference*, Warsaw, Poland, 1986, pp. 34–42.

GARFIELD – an online information system on grassy airfields

6

6.1 INTRODUCTION

Grassy airfields were the first aerodromes for airplanes. Over the years, grassy airfields have been gradually replaced by paved runways, but some are still available for small, private airplanes and gliders; for light transport aircraft in less-populated regions; and for use as emergency airstrips. General and business aviation use of aerodromes with grass runways is becoming more important (Berster et al., 2011; Reynolds-Feighan and Maclay, 2016), and there are even certain types of business or fighter jets that were designed to operate from grass [1, 2]. One drawback of the grass surface, however, is that this kind of runway is very sensitive to weather conditions, which influences trafficability of landing gear wheels and can lead to problems during takeoff and landing.

This chapter presents a project called "GARFIELD" (acronym of *G*reen, *A*ccessible and Safe European G*r*assy Ai*rfields*) focused on grassy airfields. The main purpose of the proposed project is to develop, test, verify, and introduce an online information system for aerodromes with grass runways (grassy airfields), providing data on surface conditions and on airplane take-off and landing performance. The system will be available first for grassy airfields throughout the European Union (EU), but it will be expandable all over the world. The project is an initiative of seven European universities of Denmark, Germany, Italy, Poland, Sweden, and the United Kingdom as well as five enterprises that are active in aviation.

DOI: 10.1201/9781003312765-6

6.2 MOTIVATION

Generally, grassy airfields have poorer parameters for wheel–surface performance than paved runways. Most problems are usually caused by weather and linked wheeling effects. A very important factor, however, is that there are no official guidelines on how to use grassy airfields in relation to airfield performance and weather impact; no equivalence to the International Civil Aviation Organization (ICAO) braking action estimations for hard runways exists for grassy runways. The effect of soil–grass deformability and high sensitivity to climatic conditions make airfield performance on a grass runway more difficult to predict. Table 6.1 compares several information systems for grassy airfields; those closest to the proposed "GARFIELD" system are described in the table.

In the United Kingdom, the Light Aviation Airports Study Group (LAASG) was established as a direct initiative arising from the Civil Aviation Authority (CAA)–Industry Joint Review Team (JRT) in early 2005 (LAASG, 2005). According to LAASG, the runway surface condition is very important and should be kept as smooth and well drained as possible; a well compacted and drained grass surface will be usable by light aircraft in all but abnormally wet conditions. The recommendation is that grass be kept to a maximum of 10 cm (4 inch) high. A surface can be regarded as acceptably smooth if a 1.5–2.0 Mg truck can be driven over it at 30 mph without undue discomfort to the driver. The runway must be capable of supporting the heaviest aircraft likely to use it [3].

In the United States, a research project called Opportune Landing Sites (OLS) has been conducted to collect surface data at a number of places suitable for military aviation operations (Shoop et al., 2008). Mapping software has been developed that uses commercially available Landsat images to remotely locate unimproved landing sites that are sufficiently flat and free of heavy vegetation, obstacles, and surface water to allow airlift operations when soil and weather conditions permit. The system can determine the soil type on the basis of the pixelated satellite imagery and digital terrain elevation data. Finally, the soil Moisture Content (MC) is determined, and the software infers the California Bearing Ratio (CBR) data for a given spot [4].

The Lublin University of Technology developed a test method to evaluate grassy airfields by determining the mechanical properties that affect airfield performance during takeoff, touchdown, and landing roll (Pytka, 2014; Pytka et al., 2017). The method consists of measuring equipment, a portable tester, software, a mobile application that can determine airfield performance of a given airplane, and an online application that can predict physical properties of the surface with respect to weather impact. The method was calibrated and validated in field tests with the use of a General Aviation (GA) airplane [5, 6].

TABLE 6.1 Comparison of selected grassy airfield information systems

SYSTEM OR INITIATIVE	*TYPE OF SERVICE*	*COVERAGE*	*FORE CASTING*	*PROVIDED DATA*	*NOTICE TO AIRMEN*
LASSG	Recommendations	The United Kingdom	No	General information on registered airfields	N.a.
NAVIGEO	Dynamic map	France	No	General information on registered airfields, navigation aids	N.a.
Avioportolano	Static map	Italy, Malta, Corsica, Slovenia, B&H	No	General information on registered airfields, navigation aids	N.a.
OLS	Dynamic maps	The USA	Yes	Surface CBR and roughness	N.a.
GARFIELD	Dynamic maps	Europe	Yes	Detailed information on ground performance	GRASS TAM

These initiatives show a strong societal interest and need for better solutions that would help to more precisely predict the long-term and safe use of grassy airfields. Piloting magazines publish accident analyses or training articles very frequently (Garrison, 2013; Hirschman, 2014; Landsberg, 2016, 2017; Mauch and Kutschke, 2016). It is clear that GA and the private aviation sector suffer from lack of a system such as that proposed by the authors [7–10, 11].

None of the currently available solutions provide online information on grassy runways in a form that can be operationally usable (see Table 6.1). Moreover, there is no known solution in a coupled soil–weather–wheel–airplane model that can be simply adopted or applied to the proposed system.

6.3 GARFIELD SYSTEM DESCRIPTION

The "GARFIELD" system will provide conditions of grassy runways by determining their soil mechanical properties together with grass as a function of weather conditions that affect aircraft performance during takeoff, touchdown, and landing roll.

At the heart of the system is a method to determine wheel–surface interaction parameters (rolling resistance and surface friction) on the basis of a simulation model that links meteorological conditions with soil conditions on a given airfield. The model provides simple, readable information about current conditions on a given grassy airfield anywhere in Europe. The information will be provided to end-users (pilots, air traffic controllers, airfield administrators) with a 1-hour refresh rate, together with a 6-hour forecast. The following information will be available:

- rolling resistance and braking friction or braking action for a typical GA aircraft tire;
- takeoff and landing performance data at a given airfield, specific for each aircraft category potentially using the site; and
- runway condition information in coded format (tentatively called GRASSTAM, similar to aviation NOTAMs).

The "GARFIELD" data will be updated hourly and made available in two formats:

- text format to be used by mobile applications and for desktop/laptop applications and
- coded format, provided in the form of a notice to airmen (GRASSTAM).

6.4 METHODOLOGY

The development of the system and its validation and calibration are based on some test and research methods, simulation models, and specific instrumentation, which all are described briefly in the following subsections.

6.4.1 Wheel–grass interaction modeling

Wheel–grass interaction modeling is the basic element of the system, since the performance of an airplane on a grass runway is affected by wheel performance and consequently the wheel–grass interaction. A terramechanical model has been developed taking into account the effect of speed on soil deformation $\varepsilon(t)$, soil nonlinearities, and rheological properties quantified by means of two moduli of elasticity E_1 and E_2 with $\sigma(t)$, soil mechanical stress, and A and B as constants. The complete model is expressed in the following differential equation for soil dynamic response [12]:

$$E_1 E_2 \varepsilon(t) + (E_1 + E_2)\frac{A\sigma(t)}{\sinh(B\sigma(t))}\frac{d\varepsilon}{dt} = E_1\sigma(t) + \frac{A\sigma(t)}{\sinh(B\sigma(t))}\frac{d\sigma}{dt} \quad (6.1)$$

From Eq. (6.1), a stress–strain curve is obtained for the soil under the load transmitted by the undercarriage wheel. Then, using the Bekker's formula [13], the rolling resistance and the braking friction force are determined:

$$F_{BR} = b\int \tau \, dx \quad (6.2)$$

$$F_{RR} = b\int \sigma \, dz \quad (6.3)$$

Model parameterization is possible on the basis of tabular data contained in soil maps. Knowing the location of a given airport, the type of soil is identified, and then, using tables, soil parameters are adopted. Automatic search will be facilitated by a computer database of grassy airports.

The effect of humidity is taken into account using the Anderson model [14], and the effect of vegetation is quantified on the basis of sampling at given time intervals (e.g., every week).

6.4.2 GARFIELD software development

The most important product of the "GARFIELD" project is a desktop application that will be the user's tool to communicate with a server and to get required information on given runway conditions and airplane performance. In its primary form, the desktop application shows predicted Rating Cone Index (RCI) value for a given airfield. Generally, the architecture of the projected system for grassy airfields installs the memory-consuming elements (aircraft database, soil database) as well as the simulation programs on the "GARFIELD" server, while the mobile applications ask for the required data, receive it, and display it on a smartphone or laptop computer. All communications is possible by today's wireless technology, and online access will be provided for potential end users.

6.4.3 System validation

The validation of the complete system is done by airfield measurements and flight testing. The system is verified by doing a simple comparison between actual measured performance and that determined by the airfield performance model takeoff and landing distances. To represent different surface conditions, flight testing will be performed through the entire season.

Calibration of the system requires periodic measurements of critical surface parameters and comparisons of those measurements with system predictions. A portable surface tester has been developed as a support tool to calibrate the system [15].

6.5 GRASSTAM – A NOTICE ON GRASS RUNWAY SURFACE CONDITION

6.5.1 Introduction

An important function of air traffic service is to provide information to airmen about important circumstances that affect navigation, flight planning, and flight operations. Various systems and tools are used for this purpose, among them is the NOTAM (*NOtice To AirMen*), which is a notice containing information concerning the establishment, condition, or change in any aeronautical facility, service, procedure, or hazard, which may affect flight operations [2]. NOTAM's purpose is to inform pilots of changes to airports, airways, and local procedures that can affect safety. NOTAMs are encoded and are communicated using the fastest available means to all addressees for whom the information is assessed as being of direct operational significance, and who would not otherwise have at least 7 days' prior notification. One of the types of NOTAM messages is SNOWTAM, which concerns icing or snow contamination on the runway and is based on the values of the μ coefficient, measured using methods available at a given airport, for example, a friction tester. The braking action assessment can also be based on the subjective rating provided by the crew of the landing plane (PIREP – Pilot Report). Another type of NOTAM is the ***Field Condition*** (FICON) notice [16], which describes physical condition of a runway in the sense of braking friction. The FICON is based on μ readings and surface contamination description. Those data are the base to form a runway assessment matrix (RCAM). A computer program generates a Runway Condition Code (RCC) related to the slipperiness of the surface. RCCs are only used on paved runways, therefore the current FICON system does not cover aerodromes with grass runways [17].

6.5.2 An idea of the GRASSTAM

The idea of the present author was to introduce a specific type of NOTAM, related to grass runway conditions. This new notice, named GRASSTAM, would provide information on the suitability of the runway grass surface at a given airport for air operations, on the basis of the following factors: soil mechanical strength, soil MC, vegetation, and weather conditions. By design, the GRASSTAM would be issued for ICAO-coded aerodromes with grass runways. The purpose of the GRASSTAM is to give possibly accurate field conditions, which enable to calculate the resulting airplane performance at takeoff or landing.

6.5.3 Method of grass runway assessment

The assessment of the grassy runway would be done on the basis of the following factors:

- area mass of green vegetation,
- CI,
- soil MC,
- current weather on the airfield, and
- braking friction and rolling resistance coefficients.

The evaluation of the surface conditions of a grass runway is based on the assessment matrix in which important criteria are collected. A sample assessment matrix is shown in Figure 6.1. The idea of this matrix is based on the Runway Condition Assessment Matrix [16]. The evaluation criteria proposed in the table are indicative and based on the results of the authors' own research, included in Chapter 5.

The first criterion is the parameter characterizing green vegetation. In the original version of GRASSTAM, vegetation was characterized by means of grass blades' height [18, 19]. For this purpose, a special sensor was designed, which operated on the principle of an optoelectronic sensor – a photocell. Based on the results of own research and on the basis of the analysis of the results of Shoop et al. (2015), it was found that the optimal parameter characterizing green vegetation on the runway is the mass of the green parts of the grass per unit surface area, m_G. This is indicated by a correlation of the wheel traction parameters and the m_G [20], as well as the relatively easy method of determining the unit weight of vegetation. The range of m_G values is from approximately 0.3 kg/m^2 for short dry grass, through approximately 0.45–0.6 kg/m^2 for a typical grass strip with mowed vegetation, to approximately 1.5–1.8 kg/m^2 for lush, tall grass. The impact of vegetation on the assessment is negative, which means that the higher the m_G parameter, the lower the rating of a given runway.

Another criterion is soil MC. This parameter is strongly correlated with the mechanical strength of the soil and can be determined (measured) with any moisture meter. In this project, MC was measured using a handheld Time Domain Reflectometry (TDR) meter.

Weather conditions are available from many weather services and can be used on a subscription basis. Weather conditions can have both positive and negative effects. If the weather is sunny, the air temperature is high, and the wind blows which causes the grass and soil to dry out, then it is possible to raise the score due to the favorable prognosis for the near future conditions. It is different when the weather is rainy, cool, and cloudy, then the influence of the weather is neutral (soil and plant conditions will not change in the next few hours) or negative (conditions will worsen).

The CI is a parameter that captures several soil characteristics, including cohesion and shear strength. In the original version of GRASSTAM, the ground evaluation parameter was CBR; however, due to the rather complicated procedure and measuring equipment, it was abandoned, and CI was used instead of CBR. The measurement of this index is easy, and the device is simple and relatively cheap.

Braking friction coefficient is a measurable parameter, but a typical μ measurement procedure requires the use of a friction tester, which is standard equipment at communication airports. For grassy runway airports where the

GRASS RUNWAY CONDITION ASSESSMENT MATRIX (GRCAM)						
ASSESSMENT CRITERIA				DOWNGRADE CRITERIA		
GREEN VEGETATION	**SOIL MC**	**WEATHER**	**CODE**	**μ**	**CL**	**PIREP**
Sparse < 0.25 kg/m^2	Low < 10%	CAVOK w >5 m/s t > 20°C	VI	Very good > 0.5	Very high strength CI > 3,000 kPa	Pilots reporting on: - T-O and landing distances; - directional control; - wheel sinkage; - braking deceleration; - surface roughness/ vertical vibrations
Low 0.25–0.5 kg/m^2	Low/medium 10–15%	4/8 Oct w = (3–5 m/s) t = (17–20°C)	V	Good 0.4–0.5	High strength CI = (1,800–3,000 kPa)	
Medium 0.5–0.8 kg/m^2	Medium 15–20%	6/8 Oct w = (1.3 m/s) t = (14.17 °C)	IV	Medium 0.25–0.4	Medium strength CI = (1,200–1,800 kPa)	
High 0.8–1.25 kg/m^2	Medium/high 20–30%	Light rain t = (8–14°C)	III	Satisfactory 0.15–0.25	Medium strength CI = (1,200–1,800 kPa)	
Intense 1.25–2.0 kg/m^2	High > 35%	Rain t = (7–16°C)	II	Poor 0.10–0.15	Low strength CI = (700–1,200 kPa)	
		Heavy rain t = (5–10°C)	I	Unsatisfactory < 0.10	Ultra-low strength CI < 700 kPa	

FIGURE 6.1 Grass Runway Condition Assessment Matrix.

purchase and maintenance of a typical friction tester may be a financial problem, it is proposed to use the portable wheel tester as described in Chapter 4 of this book.

The CI value as well as the braking friction coefficient have a positive influence on the evaluation.

6.6 CONCLUSION

The online information system on grassy airfields, called "GARFIELD", has been proposed as an initiative for the GA community. The system is based upon soil data, supported by online weather service, and driven by a terramechanical wheel–grass model. Its output – text or coded information on a given grassy airfield together with takeoff or landing performance of a chosen airplane – is available online. The impact of "GARFIELD" includes improved safety of operations, reduced costs of operations in moderate or bad conditions, a wider network of grassy airfields accessible to more aircraft, better access to remote locations with no road system, and a lengthened flying season.

A new notice for airmen has been proposed, called GRASSTAM, as an abbreviation referring to information about the condition of grassy airfields. The notice will be published for selected airports with a grass runway. The idea of the notice lies in the provision of data on the condition of the airport surface in terms of the load capacity of the surface, braking friction, and additional rolling resistance. All these data result from the need to calculate the length of the takeoff or landing for a given aircraft. It is anticipated that the GRASSTAM notice may contribute to the wider use of airports with a grassy runway as well as to increase the safety level of flight operations at such airports.

REFERENCES

1. Berster, P., Gelhausen, M.C., Wilken, D., 2011. Business aviation in Germany: An empirical and model-based analysis. *Journal of Air Transport Management*, 17, 354–359
2. Reynolds-Feighan, A., Maclay, P., 2016. Accessibility and attractiveness of European airports: A simple small community perspective. *Journal of Air Transport Management*, 12, 313–323
3. CAA, 2005. *Report on the Light Aviation Airports Study Group*, Civil Aviation Authorities, UK (LAASG)

4. Shoop, S.A., Diemand. D., Wieder, W.L., Mason, G., Seman, P.M., 2008. *Opportune Landing Sites Program*, Technical Report, ERDC/CRREL TR-08–17
5. Pytka, J., 2014. Identification of rolling resistance coefficients for aircraft tires on unsurfaced airfields. *Journal of Aircraft*, 51(2), 353–360
6. Pytka, J., Tarkowski, P., Budzyński, P., Józwik, J., 2017. Method for testing and evaluation of grassy runway surface. *Journal of Aircraft*, 54(1), 229–234.
7. Garrison, P., 2013. Sunday drive. *Flying Magazine*, 8, 40–42
8. Hirschman, D., 2014. No runway? No problem. *AOPA Pilot*, 3, 58–65.
9. Landsberg, B., 2016. Soft field, soft thinking. Who is responsible for a takeoff gone wrong? *AOPA Pilot*, 9, 20
10. Landsberg, B., 2017. Margins to live by. Beautiful places, challenging runways. *AOPA Pilot*, 8, 80–84
11. Takeoff and Landing, 2011. *Good Aviation Practices Series*, Civil Aviation Authority of New Zealand, Wellington, January
12. Pytka, J.A., 2013. *Dynamics of Wheel-Soil Systems. A Soil Stress and Deformation State Based Approach*, Taylor & Francis, Boca Raton, FL, Chap. 6
13. Bekker, M.G., 1961. *Theory of Land Locomotion. The Mechanics of Vehicle Mobility*. University of Michigan Press, Ann Arbor, MI
14. Anderson, M.G., 1983. On the applicability of soil water finite difference models to operational trafficability models. *Journal of Terramechanics*, 20(3/4), 139–152
15. Pytka, J., Budzynski, P., Tarkowski, P., Piaskowski, M., 2016. A portable wheel tester for tyre-road friction and rolling resistance determination. *IOP Conference Series: Materials Science and Engineering*. doi:10.1088/1757-899X/148/1/012025
16. I.C.A.O. Annex 11, Air Traffic Service
17. N JO 7930.107 Notice on Field Condition (FICON) Reporting. U.S. Department of Transportation, Federal Aviation
18. Pytka, J., Budzyński, P., Józwik, J., Łyszczyk, T., Gnapowski, E., Laskowski, J., 2019. GARFIELD information system – old problems and new perspectives. In *Proceedings of the 6th International Workshop on Metrology for Aerospace (MetroAeroSpace)*, Torino, Italy, 19–21 June 2019
19. Pytka, J., Budzyński, P., Józwik, J., Łyszczyk, T., Gnapowski, E., Laskowski, J., 2019. GRASSTAM – an idea of a notice on grassy runway condition. In *Proceedings of the 6th International Workshop on Metrology for Aerospace (MetroAeroSpace)*, Torino, Italy, 19–21 June 2019
20. Shoop, S.A., Coutermarsh, B., Cary, T., Howard, H., 2015. Quantifying vegetation biomass impacts on vehicle mobility. *Journal of Terramechanics*, 61, 63–76

Index